Urban Warfare in Iraq 2003-2006

By J. Stevens

Urban Warfare in Iraq 2003-2006

Dedicated to the American servicemen and women who have lost their lives and limbs fighting in Iraq.

Front cover photo by SSG James Harper Jr.

Rear cover photo by LCPL Christopher Zahn.

© 2006 SSI

Colon, NC

All rights reserved.

Published in the United States of America.

No part of this book may be reproduced in any form or by any electronic or mechanical means, including information storage and retrieval devices, without prior written permission from the publisher.

Author: J. Stevens

ISBN: 978-0-9714133-6-8

Library of Congress Control Number: 2006925162

Urban Warfare in Iraq 2003-2006

Urban Warfare in Iraq provides an overview of the urban aspect of the war in Iraq from the start of hostilities in March 2003 until the date of publication in April 2006. This work conveys the war's central elements so the reader can better understand the nature of the conflict in Iraq. The format is picture based and presented in such a way as to be easily digested.

The war in Iraq is largely an urban war, one with the strongest areas of insurgent resistance centered around such urban areas as the cities of Ramadi, Fallujah, Baghdad, and Mosul. The urban areas are the centers of contention because the cities hold what is most valuable: the seat of government, financial institutions, religious centers, national infrastructure, international businesses and organizations, military and intelligence headquarters, and the majority of the nation's people.

The urban war in Iraq has mostly been low-intensity, with the understanding that there have been high-intensity exceptions like the fighting in Fallujah, Karbala, Najaf, and Sadr City in 2004. Not only is the war largely an urban one, but its central characteristic is defined by a foreign, high-intensity war machine fighting a lightly armed, indigenous guerrilla movement.

Table of Contents

1. **Urban Warfare Fundamentals** (Pages 7-34)
   Examines what defines urban warfare in Iraq, what are its essential characteristics, and how these elements affect combat operations.

2. **Insurgent Tactics** (Pages 35-76)
   Looks at the insurgents' methodology and how they wage their urban guerrilla war against government security forces.

3. **People: The Key Terrain** (Pages 77-96)
   Analyzes how the people affect urban combat operations and how they are the pivotal element in a guerrilla war.

4. **Weapons of War** (Pages 97-112)
   A critical examination of the weapons of war used in Iraq and their advantages/disadvantages.

5. **Aircraft in the Cities** (Pages 113-124)
   Assesses the capabilities and limitations of airpower in an urban environment.

6. **Close Quarters Battle (CQB)** (Pages 125-172)
   An in-depth look at CQB fundamentals to include movement, elevation, breaching, climbing, stairways, hallways, urban camouflage, windows, lighting, and common mistakes. This chapter also looks at the fundamentals of urban sniping as it is employed by both insurgents and American forces.

War in Iraq Timeline

March 19, 2003 - United States invades Iraq.

March 29, 2003 – First suicide car bomb attack against U.S. forces.

April 9, 2003 – Baghdad falls.

April 14, 2003 – Saddam Hussein's hometown of Tikrit falls.

May 1, 2003 - Official end of hostilities.

July 22, 2003 – Uday and Qusay Hussein are killed in Mosul.

August 7, 2003 – Jordanian embassy in Baghdad is car bombed.

August 19, 2003 – UN HQs in Baghdad is car bombed.

October 27, 2003 – Red Cross headquarters in Baghdad is attacked with an ambulance car bomb.

December 13, 2003 - Saddam Hussein is captured at a farm in Tikrit.

January 18, 2004 – U.S. military HQs in Baghdad is car bombed.

March 31, 2004 – 4 Blackwater employees killed in Fallujah.

June 28, 2004 – Authority is transferred to the Iraqi government.

October 24, 2004 – Abu Musab Zarqawi's group executes 50 recruits.

November 8, 2004 – American forces begin assault to retake Fallujah.

January 30, 2005 – Elections for 275-seat National Assembly held.

February 28, 2005 – Suicide car bomber kills 115 people in Hillah.

October 2005 – Saddam Hussein trial begins.

October 15, 2005 – Millions of Iraqis vote on new constitution.

October 26, 2005 – 2,000th U.S. soldier is killed in IED attack.

December 15, 2005 – Iraq holds parliamentary elections.

February 21, 2006 – Shiite shrine in Samarra is blown up, sparking widespread sectarian violence.

March 7, 2006 – Gunmen kidnap 50 employees of an Iraqi security company.

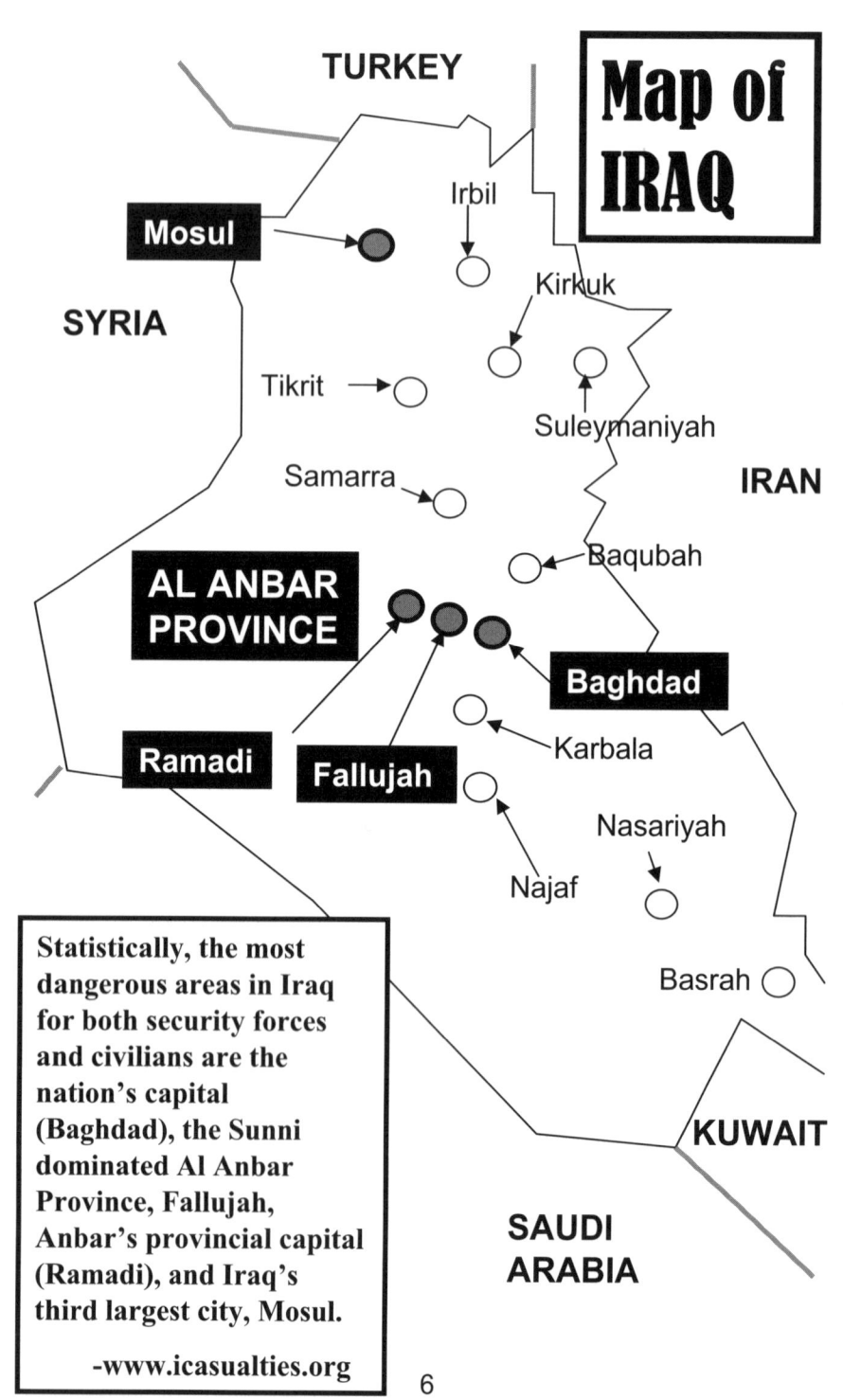

Chapter 1
Urban Warfare Fundamentals

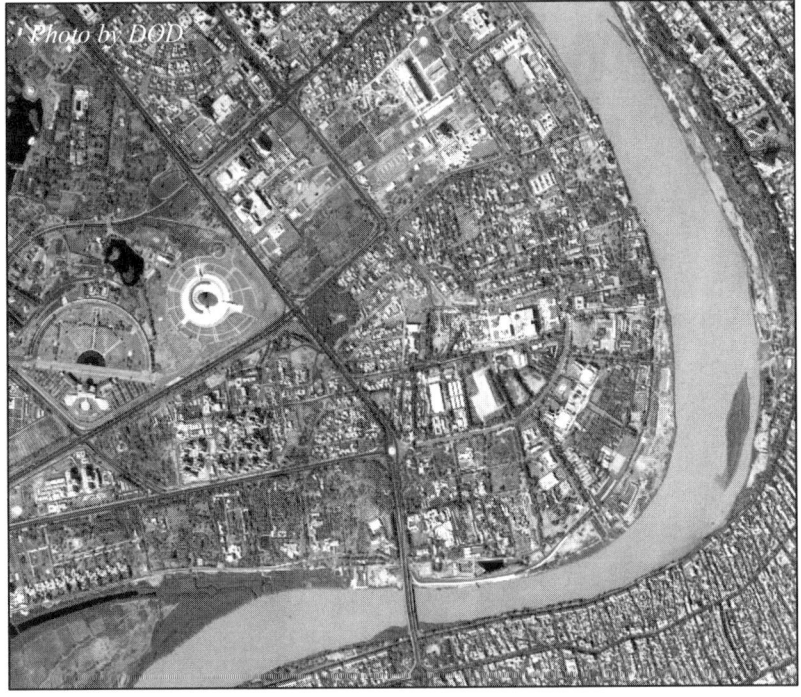

(An aerial photo of Baghdad's downtown area to include the "Green Zone" subsequently renamed the "International Zone".)

Iraq's capital of Baghdad is the nation's deadliest city – leading in number of insurgents, attacks, kidnappings, security force casualties, civilian deaths, and homicides.

Chapter 1 – Urban Warfare Fundamentals

Urban Sprawl = A Complex Environment

Iraq's major cities like Baghdad, Mosul, Basrah, Karbala and Najaf are sprawling areas with millions of inhabitants. These millions of people have different religious, political, ethnic, tribal, and national loyalties of varying degrees. Different neighborhoods hold different loyalties. A single street may separate different ethnicities or nationalities. People on the top floor of an apartment building may have an entirely different political view than the people on the bottom floor. Variety equals complexity.

Because of this complexity, assimilating these different elements into a coherent framework is a monumental task. Most military organizations can only view a city in the terms of general trends because they cannot absorb the sheer volume of the city.

The cities' human complexities are matched by their physical complexities with millions of individual buildings, structures, and homes. The urban areas are composed of an endless variety of obstacles, both human and structural, to include bridges, overpasses, potholes, traffic jams, crowds of people, power lines, culverts, collapsed buildings, and dangerous slums.

Chapter 1 – Urban Warfare Fundamentals

Traffic is a Central Part of the Environment

Photo by Mathew Accosta

As this line of cars in Tikrit shows, Iraqi society is a motorized one. There are millions of cars in Iraq, many of them sold illegally on the black market. Cars are cheap and so is gas. Everywhere in the cities there are traffic jams, honking horns, car accidents, angry motorists, and rampant pollution. A continuing problem is the country's decaying infrastructure, where crumbling roads, potholes, and blast craters impede traffic and cause high wear and tear on vehicles.

Heavy traffic and traffic jams caused by government checkpoints, regular car accidents, and IEDs become obstacles to movement in the urban environment. These vehicular obstacles impede the movement of government security forces, often bringing their patrols to a standstill.

> **Before the March 2003 invasion there were 1.5 million registered cars. In October 2005 there were 3.1 million.**   **- Brookings Institution Iraq Index**

Chapter 1 – Urban Warfare Fundamentals

The People: The Most Important Element of the Urban Terrain

An estimated 26 Million people live in Iraq with 75% of the population (19 million people) living in the cities. Iraq's ten largest cities have 15.7 million people, or 60% of the country's total population (Baghdad: 6 million, Basrah: 2.6 million, Mosul: 1.7 million, Irbil: 1 million, Kirkuk: 750,000, Suleymaniyah: 800,000, Najaf: 600,000, Karbala: 500,000, Ramadi: 400,000, Fallujah: 350, 000).

This large urban population makes the people as much a part of the urban terrain as the surrounding buildings and roads. People are everywhere and security forces must interact with them on a daily basis. In fact, it is impossible for security forces to separate themselves completely from the people.

If the average citizen can get close to security forces in the course of their daily activities, so can the insurgents – in the form of sniper attacks, assassinations, suicide bombings and hostage-taking. In this sense, the people are the insurgents' cloak, hiding them until they come in the open to conduct attacks, and then protecting the insurgents again when they are absorbed back into the population.

Chapter 1 – Urban Warfare Fundamentals

A Rural-Centric Force = Failure

In contrast to the urban reality of buildings, cars, people, and guerrillas, most of the world's militaries are conventional and rural-centric in their doctrine, training, and equipment. Since most wars in history have been rural in nature, modern militaries are a reflection of this rural past.

The above camouflaged soldier is only effective in an environment absent of people, cars, and buildings. A conventional military force is at a distinct disadvantage in a city because wearing a uniform equates to an inability to blend into the urban environment.

This inability to assimilate into their surroundings is contrasted by the insurgents, who take great efforts to blend into their environment and are able to operate and survive against overwhelming odds due precisely to this assimilation mismatch.

Chapter 1 – Urban Warfare Fundamentals

Overcentralization = Paralysis

Photo by Martin Greeson

Conventionally grounded commanders only comfortable moving large formations of men and material around on a battlefield will not be successful. In the cities, life is fast, quick decisions outside of the centralized chain of command must be made, and junior leaders must make on the spot decisions.

Commanders must develop and then trust their subordinate leadership to execute the commander's intent and campaign plan in a decentralized manner. Overcentralization, combined with its resulting loss of initiative, equals operational paralysis.

Another reason why urban guerrillas can survive under the scrutiny of large numbers of conventional security forces is that the insurgents have a superior decision making cycle. Simply put, because the insurgents' operations are truly decentralized they can make operational decisions and act faster than their relatively clumsy opponents.

Chapter 1 – Urban Warfare Fundamentals

Small Unit Leadership is Key

The urban battlefield forces units to operate in small groups, at the team and squad level, whether they want to or not. Overly centralized organizations will have problems when the urban terrain makes them become decentralized against their will.

Successful combat organizations have aggressive, confident NCOs and platoon leaders who understand their commander's intent and then accomplish the mission as they see fit. Small units, like the one depicted above, are the units that fight the war in the urban battlefield. Hence, urban wars are won or lost depending on the quality of these small-size elements.

Fire teams, squads, and platoons cannot be micromanaged successfully in the urban environment as they can be in the simpler, rural battlefield. Only aggressive, small-unit leadership will prove successful in the physically complex cities.

Chapter 1 – Urban Warfare Fundamentals

Operational Compression

Many things are compressed in a city – living space, roads, buildings, and military organizations. Military organizations that normally control large tracts of land in a rural setting may only control a single neighborhood in the cities. However, in this neighborhood there may be a population of one hundred thousand people with a small, invisible insurgent organization hiding among these people.

Because of this operational compression, different units may have to coordinate at the brigade level merely to conduct a squad-size raid across the street, but a street that happens to be a unit boundary. This unusually frequent need to deconflict unit boundaries is unique to the urban environment.

The insurgents exploit these myriad boundaries intentionally, hopping them to confuse security forces. Successful organizations will not let the scores of operational boundaries in the compressed urban environment reduce their operational flexibility.

Chapter 1 – Urban Warfare Fundamentals

High Intensity Warfare Methods Are Limited in a Guerrilla War

Photo by DOD

A military's high-intensity warfare machine can only be unleashed if infrastructure damage and civilian casualties are not a concern. The American military's operations in Fallujah in November 2004 is an illustration of such warfare, where the U.S. military operated with few restrictions. However, this type of all out warfare is the exception in Iraq.

In contrast to the pitched battles in Fallujah, low intensity guerrilla warfare is the norm and is the defining character of the war. In this guerrilla war human intelligence, precision weapons systems, information operations, civil affairs, infrastructure efforts, special operations, and a sound detention/interrogation/legal system are more important than massed forces, heavy firepower, and operational mobility. Successful organizations will modify their tactics to fit the environment while unsuccessful ones will try and force their existing methodology on the urban terrain.

Chapter 1 – Urban Warfare Fundamentals

Intelligence Collection

(An American soldier launches an unmanned aerial drone.)

Collecting intelligence in the urban world is difficult due to its complexities, architecture, and infrastructure. However, a variety of aerial platforms to include remote planes, helicopters, aircraft, and satellites can observe real-time street activity. The bird's-eye view feature of aerial platforms allows them to overcome the visual screening that impedes a person looking at a city from street level. These intelligence collection platforms work better in a rural-centric environment and less so in the cities.

Aerial platforms are limited in their effectiveness because the cities are a three dimensional world. Top-down looking platforms can only see what is on the streets and nothing else. These platforms cannot see through buildings, cannot look through roofs, cannot peer into cars, and cannot penetrate into basements and underground sewers and tunnels. Electronic means of intelligence collection only provide limited success when the most important terrain is a human one. It is easy for technology dependent organizations to over-focus on technical means of intelligence collection as opposed to working with the local populace to get the intelligence required to penetrate guerrilla organizations.

Chapter 1 – Urban Warfare Fundamentals

The Phenomena of Microenvironments

Urbanization results in an uncountable number of microenvironments existing in every city. The Fallujah apartment building shown above has its own environment inside. The scores of individual apartments inside the building each have their own individual environments. Every separate room in each one of the separate apartments also has their own environment. A single building like the one above may have one hundred separate microenvironments, each one isolated to a certain degree from the others.

As a result, someone on the fourth floor may be making an IED and the person next door to him will not know this, never mind the person living on the bottom floor. The Marines patrolling down on the street are operating in a separate environment and certainly cannot see or hear what is happening on the fourth floor. The sheer volume of microenvironments contribute to making the urban environment a complex one and individual soldiers only have partial awareness of the entire environment at best.

Chapter 1 – Urban Warfare Fundamentals

Infantry is King

Photo by CPL Mike Escobar

Because of three dimensional microenvironments like an apartment building, mechanized forces cannot operate independently in an urban environment and succeed. People in vehicles are trapped in their own microenvironment and have very little situational awareness outside of their own vehicle. History has repeatedly shown that mechanized forces fighting independently in a city receive disproportionate losses.

It is the dismounted soldier that has the greatest urban mobility. The foot soldier can walk up a flight of stairs, sneak through back alleys, climb through windows, scale buildings, travel from rooftop to rooftop, slither through sewers, and search buildings room by room – vehicles cannot.

Importantly, the dismounted soldier uses their superior situational awareness to protect mechanized forces that are vulnerable in the built-up environment. Soldiers on foot can talk to the people while soldiers in a vehicle cannot establish relationships or receive tips on local guerrilla activity as they speed by in their enclosed microenvironments.

Chapter 1 – Urban Warfare Fundamentals

The Problem of Relative Isolation

Photo by CPL Mike Escobar

Operating in microenvironments means soldiers will suffer repeatedly from the problem of relative isolation. As soon as the two Iraqi soldiers shown in the above picture walked down the stairwell, they lost visual and verbal contact with the other members of their patrol. This relative isolation makes communications very difficult between separated elements.

These two soldiers may see something below them and not be able to communicate it to the rest of their patrol. Or, these two men could be captured or killed and the rest of their patrol could be oblivious to their plight unless the patrol hears something out of the ordinary. There is no solving this problem since the urban terrain physically disrupts all forces working in it.

However, organizations can prepare for this disruption by training in situations where they become isolated. Training organizations and junior leaders to regularly operate in small-size elements is the best way to minimize the disruptive phenomenon of relative isolation.

Chapter 1 – Urban Warfare Fundamentals

Fighting Takes Place at Close Ranges

Photo by SSG James Harper Jr

In urban warfare combat often takes place at close distances. The soldier shown above has to be aware of threats that may be hidden behind the door and around the corner. A street only twenty-five meters wide may separate a row of buildings occupied by security forces and another row of buildings occupied by the enemy.

Twenty-five meters may actually be considered far when compared to other situations. Think about security forces clearing an apartment building floor by floor, room by room as they look for a hidden enemy. These security forces may then have to fight it out with guerrillas that are only located ten feet away in an adjacent room or across a hallway.

Once in a room, a soldier may turn a corner and come face to face with their foe. A soldier may open a door and be standing toe-to-toe with their enemy. In these situations, enemies may fight it out only several feet from each other, shooting at each other through walls, floors, doors, windows, and ceilings. Entire gunfights can occur without either side actually seeing their enemy.

Chapter 1 – Urban Warfare Fundamentals

Increase in Casualties

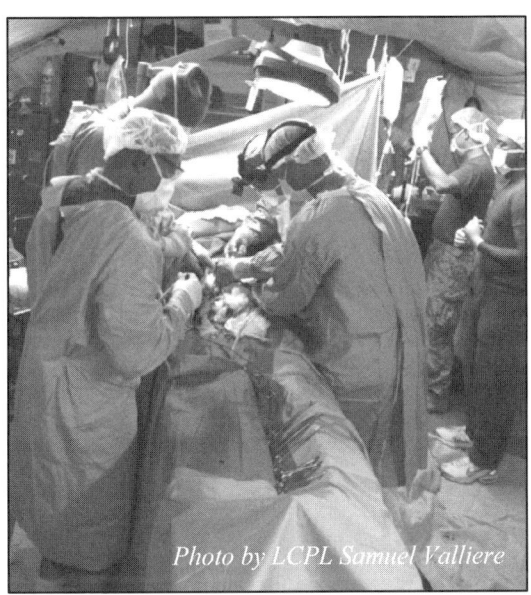
Photo by LCPL Samuel Valliere

Fighting at extremely close distances results in a larger number of casualties among combatants. During high intensity urban warfare grenades, air strikes, snipers, and artillery all take a heavy toll. During urban guerrilla warfare the casualties caused by guerrillas come from car bombs, IEDs, and ambushes.

Additionally, the enclosed nature of the urban environment increases the destructive effect of weapons like hand grenades and RPGs. Walls and floors contribute to ricochets, allowing soldiers to skip rounds and shrapnel into their opponents without actually seeing them. Security forces locked in battle with enemy forces located only meters away are also more vulnerable to fratricide, because distinguishing friend and foe is sometimes impossible.

If security forces uniformly wear ceramic body armor and Kevlar helmets they will suffer from a decreased number of fatal thoracic and head injuries. However, medical personnel will see an increased number of casualties from blast injuries and wounds to unprotected extremities like arms and legs.

Chapter 1 – Urban Warfare Fundamentals

Combat Fatigue is Increased

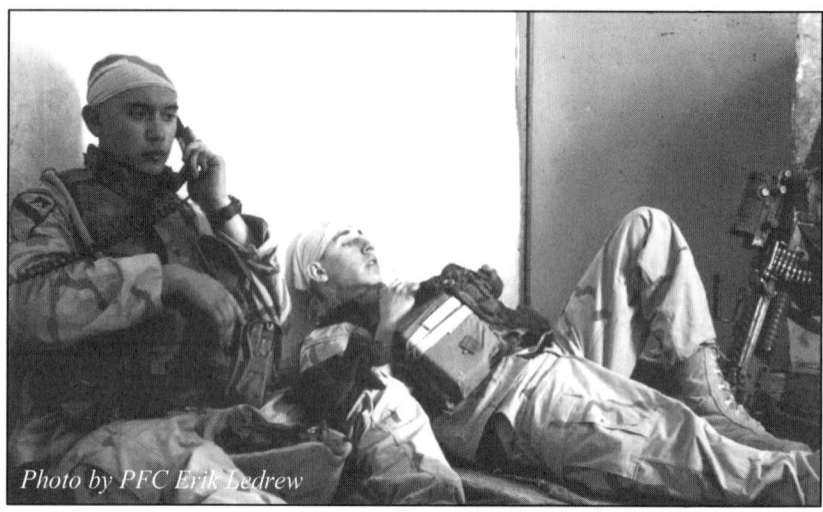
Photo by PFC Erik Ledrew

Due to the fast paced urban environment with thousands of possible threats, soldiers become fatigued faster than in a rural environment. Wearing heavy body armor and carrying large amounts of ammunition required for street fighting also increases fatigue. Additionally, the uncertainty from fighting a low-intensity guerrilla war, with a faceless enemy who refuses to fight in a conventional manner adds to the stress of the environment.

Photo by SGT Luis Agostini

22

Chapter 1 – Urban Warfare Fundamentals

Limiting Infrastructure Damage

Photo by PFC Elizabeth Erste

Precision weapon systems combined with a mentality of operational restraint and escalating force are required to minimize collateral infrastructure damage. Therefore, security forces require a broad spectrum of weapon systems that can be applied to any given situation in order to not unnecessarily destroy the cities. Also, repairing Iraq's cities are extremely costly, requiring billions of dollars just to bring it to pre-war conditions.

Every blown up building is a propaganda coup for the insurgents. Plus, ruins may become even more suitable as platforms for sniper attacks and ambushes. The insurgents often attack from partially destroyed buildings because the immediate absence of people results in a more secure operation.

The World Bank estimated it would cost over $24 billion to rebuild Iraq's damaged infrastructure.
- Brookings Institution Iraq Index

Chapter 1 – Urban Warfare Fundamentals

Rubbling Creates Problems

Photo by PFC William Servinski II

The creation of rubble through high-intensity warfare tactics has several operational consequences. First, highly rubblized areas are impassable to most military vehicles, limiting security forces' movement. Rubble, like that illustrated above can puncture tires, throw tracks, and cause vehicles to bottom out.

Rubble is also difficult to traverse by dismounted soldiers, slowing their speed considerably. Soldiers caught in fields of rubble by enemy forces have no cover and are vulnerable targets. Consequently, fields of rubble become obstacles for security forces traveling on foot, hindering them more than it does the insurgents.

Since rubble is an obstacle to the average person, the civilian populace is inconvenienced because they cannot drive or walk through it. Most importantly, rubble is an eyesore, constantly reminding the populace of the government's failures.

Chapter 1 – Urban Warfare Fundamentals

The Importance of Rebuilding

(Water pipelines being repaired in Baghdad.)

(U.S. Army personnel putting in new sewage lines.)

Limited manpower, money, and security available to make infrastructure repairs is a strong reason for pursuing a war of restraint in the cities. It does not matter if the government destroys infrastructure through aerial bombs or if the insurgents do it through car bombs. In the end, it is the people who suffer and the government ultimately has to foot the bill to restore the damaged infrastructure.

However, infrastructure improvements that better the people's health and overall standard of living is a powerful tool if wielded effectively by the government. An increased quality of life, if combined with an effective information campaign, can translate to increased legitimacy for the government.

A December 2005 poll showed that 80% of Iraqis who were asked placed infrastructure repair as one of their top three priorities.
- Brookings Institution Iraq Index

Chapter 1 – Urban Warfare Fundamentals

The Importance of Infrastructure

Photo provided to the DOD

The above picture shows a newly refurbished megawatt power plant located in the southern city of Basrah. Despite having some of the world's largest oil deposits, power shortages and rolling blackouts are common in Iraqi cities, even in cities that have functioning power plants.

This lack of a steady supply of electricity exasperates the average citizen, undermining the government's legitimacy. Providing power for the country is a basic service the government must provide to win popular support from the people. The infrastructure-popular support-government legitimacy link is another element of the urban terrain that must be taken into account by security forces.

A September 2005 poll showed that inadequate electricity impacted Iraqis' daily lives most, *not* the insurgency.
- Brookings Institution Iraq Index

Chapter 1 – Urban Warfare Fundamentals

Infrastructure: Financial Institutions

Photo by PFC Thomas Day

(Below, a Baghdad bank exchanges new dinars for old ones.)

Photo by DOD

The banking industry is a crucial component of any nation's infrastructure. The public must be confident that their money will be protected by the government and will be safe when deposited in a recognized financial institution, like the Bank of Mosul shown at the top left. Because financial stability is crucial, banks are another institution that requires government security forces to protect them and their customers.

Financial security and stability is important as it relates to the value of the Iraqi dinar. Inflation undermines consumer confidence and the person on the street gets less goods for their hard earned dinar. Also, foreign investors will not risk their money in a country that has its banks being blown up by guerrilla forces.

Iraq's 2003 inflation was at 36%, in 2004 it was at 32%, and in 2005 it was at 20%.
- Brookings Institution Iraq Index

Chapter 1 – Urban Warfare Fundamentals

Protecting The Infrastructure

Photo by DOD

Because key infrastructure like airports, banks, water treatment plants, power plants and electric grids like the one shown above are important for government legitimacy and popular support, government forces must expend precious manpower to protect them. This defensive posture, although necessary, reduces the number of security forces available for other duties like offensive operations and policing the streets.

In contrast, the insurgents do not have to protect anything. In fact, the insurgents want the government to expend as much manpower as possible in defending critical infrastructure. This tying down of troops gives the insurgents more freedom of movement on the streets, allowing them to mass their forces and attack where security forces are weakest.

> **By February 2006, insurgents had conducted 298 attacks against oil facilities and personnel.**
> **- Brookings Institution Iraq Index**

Chapter 1 – Urban Warfare Fundamentals

The Mosque System

(Soldiers stand guard in front of just one of Iraq's thousands of mosques.)

Islam is an important aspect of every day Iraqi life and its influence on the general population cannot be overstated and should not be underestimated. Iraqi Muslims stay grounded in their faith, in part, by attending religious services at local mosques that are centrally located in every community.

A large mosque system has long been established in Iraq and the greater Baghdad region alone has close to a thousand mosques. The most hostile city in Iraq, Fallujah, is known as "The City of Mosques".

**A November 2005 poll showed that religion had the greatest influence on whom Iraqis' would vote for.
- Brookings Institution Iraq Index**

Chapter 1 – Urban Warfare Fundamentals

The Mosque Powder Keg

Photo by SGT Neil Fucci

In the above photograph American Marines prepare to enter a mosque. While security forces must search mosques because they are frequently used by the Islamic-based insurgents, this picture represents a win/win situation for the insurgents.

The insurgents win no matter what the outcome because insurgent exploitation of the mosque system intentionally places American forces in a difficult situation. By American forces entering and violating the sanctity of the mosque, the insurgents secure a propaganda victory.

From the insurgent perspective, what better way to win popular support from a religiously grounded populace than showing American soldiers with boots on, carrying guns, and using dogs to force their way into a revered house of worship?

Importantly, the Islamic-based insurgents view the mosque system as a political football to be manipulated to further their cause. The insurgents castigate security forces for violating the sanctity of the mosques, knowing that the insurgents themselves have already violated these same mosques for their own movement. The winner of this mosque perception battle is whichever side that can best leverage the issue through their information operations.

Chapter 1 – Urban Warfare Fundamentals

Mosque Operations Have Political Significance

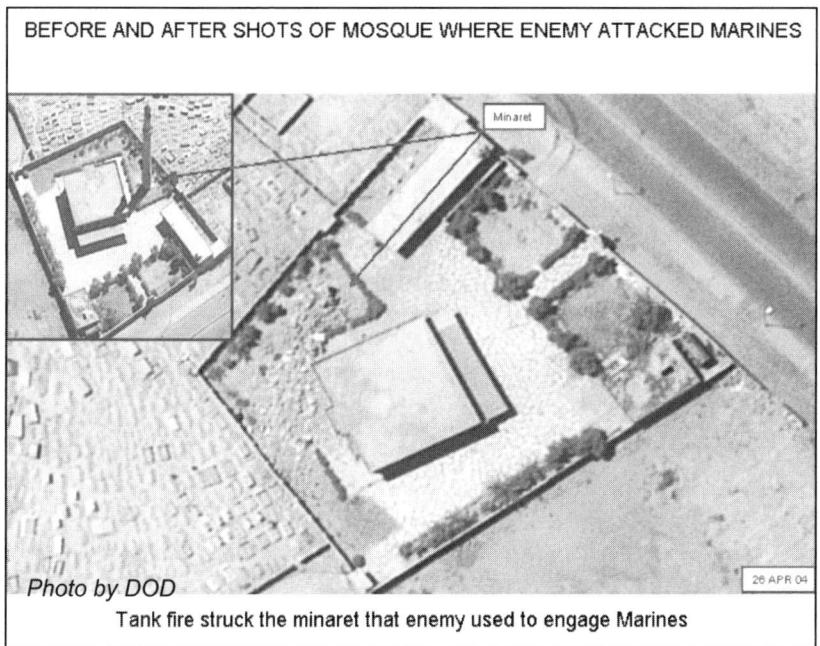

BEFORE AND AFTER SHOTS OF MOSQUE WHERE ENEMY ATTACKED MARINES

Photo by DOD

Tank fire struck the minaret that enemy used to engage Marines

An official USMC slide explains exactly how American Marines employed minimal force to destroy a mosque minaret used to attack American forces in Fallujah.

This briefing helps defeat insurgent allegations that American forces damaged the mosque for no apparent reason, that the Americans used excessive force, and that the insurgents did not use the mosque for anything other than religious services.

Religion is such a potentially explosive issue that extra measures must be taken to defuse it. It is smart to combine information operations with any activity that involves security forces attacking or searching a mosque. Only timely, accurate information can shape the population's perspective on the subject.

Chapter 1 – Urban Warfare Fundamentals

Wild Cards: Sandstorms

Photo by SSG Chad Chisholm

Iraq's cities, while being miles from the nearest body of sand, are still part of an arid country with large tracts of inhospitable desert. Many of Iraq's cities are essentially small urban islands in a large sea of sand. Consequently, people may be shopping for the latest computers at a posh shopping district in downtown Baghdad and then step outside of the mall into a raging sandstorm.

Sandstorms will halt military operations because helicopters cannot fly in them, lights cannot shine through them, and driving in them is hazardous to say the least. The cities are paralyzed during a sandstorm, forcing everyone inside until the storm is over.

Also, the talcum powder-like sand finds its way into every nook and cranny, causing excessive wear on engines, weapons, and anything mechanical.

Chapter 1 – Urban Warfare Fundamentals

There Are Exceptions to the Urban Environment

(American soldiers conduct a riverine patrol down the Tigris River in a Zodiac.)

Even though the defining characteristics of the urban environment are people, cars, roads, and buildings, every urban area is not uniform. While Baghdad is a sprawling city housing more than twenty percent of the country's entire population, the waters of the ancient Tigris River slither through the capital like a giant serpent. Because many major cities were built on waterways for reasons of commerce (Fallujah lies on the banks of the great Euphrates River) bodies of water with their accompanying bridges and sluices should be expected.

As a result, organizations conducting urban operations may still have the requirement to conduct rural–centric operations like riverine patrols in Zodiac boats. Indeed, security forces may be able to exploit the riverine/rural aspect of the environment to facilitate their urban operations.

Chapter 1 – Urban Warfare Fundamentals

The Urban Environment Does not Exist in Isolation

Photo by CPL Mike Escobar

Baghdad, Iraq's largest city, is an urban island surrounded in all directions by countless acres of farm land. The same is true for most of Iraq's cities like Ramadi, Fallujah, Mosul, Irbil, Baqubah, and Hillah to name a few.

Even in the cities, the larger urban areas surround smaller rural areas like parks, orchards, and private farms. While the urban environment dominates the cities, isolated rural areas are still present like islands in a sea of cars, buildings, and people.

The insurgents readily exploit the urban-rural seams to advance their movement, transitioning from urban-to rural operations and back again. Car bombs are built in the farms but are detonated in the capital. Insurgents may conduct a high-risk raid on a police station, and then go underground and hide in the surrounding countryside. In general, the insurgents' support network is in the countryside while focusing their combat operations in the cities.

Chapter 2
Insurgent Tactics

(Smoke rises from the Marine barracks in Beirut, October 23, 1983, after being destroyed by a suicide car bomb.)

(Photo Courtesy of USMC)

Chapter 2 – Insurgent Tactics

The Iraqi Insurgency

Photo by DOD

Above, an Iraqi T-55 tank burns as American forces drive towards Baghdad during the initial stage of the Iraqi invasion in March 2003. After the overthrow of the Saddam Hussein regime, opposition to the American-led invasion rapidly transitioned to guerrilla warfare. Because conventional means of Iraqi resistance were quickly crushed, only unconventional methods had a chance to succeed. From the time of the initial invasion, the insurgents have enlisted enough support for their cause that by April of 2006, the end of the ongoing guerrilla war was still not in sight.

The insurgency has been estimated at between 15,000 and 20,000 active fighters. By January 2006, a total of 57,490 suspected insurgents have been estimated to be killed or detained.

In January 2006 foreign fighters were estimated at up to 2,000 in strength.

In the year 2005 the insurgents conducted 34,131 separate attacks, up from the previous year that witnessed 24,496 attacks.
- **The Brookings Institution Iraq Index**

Chapter 2 – Insurgent Tactics

A Decentralized Movement

Photo by DOD

Even after the capture of Saddam Hussein, the insurgency continued its momentum. This organizational resiliency was due in part to the fact Saddam Hussein was just one face of a multi-headed insurgent hydra. The insurgency is difficult to combat precisely because it is composed of so many different elements that oppose the American occupation and the new Iraqi government.

These resistance elements include former Ba'athists, Iraqi nationalists, Sunni militants, Islamic fundamentalists, foreign fighters from across the Middle East, sporadic Shiite opposition, and international state supporters like Iran and Syria.

Chapter 2 – Insurgent Tactics

Unconventional Tactics

Photo by SPC Katherine Roth

The insurgents intentionally conduct their attacks in civilian clothes to deceive government security forces, making shoot/no-shoot situations difficult. In civilian clothes the insurgents blend in with the people, then secure hidden weapons, and conduct their attacks. After the attack, they hide their weapons and they are now indistinguishable from the surrounding population.

The more the insurgents act like a guerrilla force, the more successful they are against conventional security forces. The more the insurgents act like their conventional enemy, the less successful they are. While the insurgents' unconventional tactics have enabled them to successfully resist security forces for three years, they have been unable to transition into a force capable of overthrowing the new government.

The insurgents do not expect to defeat security forces through force of arms, but they do hope to outlast them politically. For the insurgents, the early withdrawal of militarily undefeated American forces is still a decisive political victory.

Chapter 2 – Insurgent Tactics

Weapons – AK-47

Photo by CPL Tom Sloan

In this picture a Marine holds the insurgents' most flexible urban guerrilla warfare tool – the folding stock AK-47 assault rifle. The AK-47's 7.62 x 39 mm round can penetrate car doors, windshields, and level IIIA body armor. Additionally, the AK-47 is cheaply made, widely available, legendary in its reliability, easy to maintain, and simple to use.

Because the above version has a folding stock, the rifle can be hidden beneath a jacket or under a car seat. This rifle can be maneuvered in tight urban confines like inside vehicles and stairwells.

The AK-47 is a staple product for any self-respecting black market dealer and can be bought in any city for around one hundred dollars. Iraq is awash in AK-47's and every household is allowed to have one for home protection.

Chapter 2 – Insurgent Tactics
Weapons – Machine Guns

The RPD is a Soviet made, belt-fed, extremely reliable, general purpose machine gun, firing a 7.62 x 54 mm round. The RPD is widely available on the Iraqi black market and only costs several hundred dollars.

Photo by LCPL Warren Peace

The PKM is a reliable, Soviet made 7.62 mm, belt-fed machine-gun. Equally available for people who know who to ask.

Photo by CPL Tom Sloan

Photo by DOD

The RPK is another Soviet made 7.62 mm machine-gun, except is fires from a box fed magazine. Also, affordable and available.

All of these guns are large and not easily concealed; they must be hidden in caches when not in use.

Chapter 2 – Insurgent Tactics

Weapons - RPG

Photo by 1LT Justin Colvin

The RPG is a Soviet made, anti-armor, rocket propelled grenade launcher. The RPG is an old but reliable weapon that can be used to destroy bunkers, guard towers, government police cars, unarmored trucks, armored vehicles, and helicopters. When fired in volleys at the right location, they can even destroy tanks.

The RPG is a flexible weapon because of its various warheads. The RPG can fire anti-personnel rockets, high-explosive anti-tank rockets, and armor piercing sabot rounds. It is easy to use, requiring minimal training, and is simple to maintain.

The RPG is a relatively big weapon and it is not easily concealed, although it can fit in the trunk of a car or inside a van. The RPG is readily available on the black market, is affordable, and is used in large numbers by insurgent forces when conducting ambushes and high-risk raids.

Chapter 2 – Insurgent Tactics

Weapons - Mortars

Photo by SGT Paul Mancuso

The insurgents use mortars to attack large military bases and important government controlled areas like the Green Zone in Baghdad. The insurgents also use mortars as an ethnic cleansing tool to bombard and terrorize the neighborhoods of rival social and religious groups. Since the mortar is an area weapon with limited accuracy, the insurgents generally stick to large targets in order to be successful.

Importantly, the mortar is a long-range weapon that allows the insurgents to conduct their attack anonymously, keeping their identity secure from their actual targets that may be several miles away. In the cities, a mortar can be hidden in the trunk of a car, removed, fired, put back in the trunk, and then driven away in under a minute. The mortar is often used as a harassment tool and to show the impotence of the government.

Chapter 2 – Insurgent Tactics

Weapons – SA-7

Photo by PFC James Matise

The SA-7 is a Soviet made, man portable, shoulder launched, heat seeking, anti-aircraft missile. Since this weapon is so valuable for shooting down aircraft, prices on the black market may run to several thousand dollars. SA-7's are available but are fewer in number than run of the mill weapons like AK-47s.

SA-7's are small enough that they can be transported in the trunk of a car or the back of an SUV or minivan. SA-7's can be easily carried and maneuvered while walking up stairs and passing through doorways. This means that they can be fired from any number of urban firing platforms like rooftops and balconies.

Several coalition aircraft have been shot down with insurgent SA-7's to include Apache attack helicopters, contracted transport helicopters, and even a British C-130.

Chapter 2 – Insurgent Tactics

Remote Detonating Systems

Photo by SPC Ben Brody

A key element of the insurgents' ability to conduct car bomb and IED attacks are electronic initiation systems. Seen above are explosive initiation components captured from an insurgent's residence in Sadr City, Baghdad.

Some of the components above include batteries, wiring, keyless door entry devices, fuses, and electric relay boxes. Collectively, these components allow the insurgents to create remote detonation systems. All the insurgents have to do is press a button and an Improvised Explosive Device will detonate.

Since the construction of remote detonation devices is such a technical skill, insurgent IED and car bomb builders rank high on the security forces' most wanted list. The insurgents strive to train as many people as they can in this skill so that the loss of one or more detonation specialists does not disrupt the organization.

Chapter 2 – Insurgent Tactics

Car Bombs

Photo by LCPL Brian Henner

The car bomb is an ideal urban guerrilla weapon because the vehicle itself provides concealment for its payload, leaving the explosives undetected to the naked eye. Even a relatively small sedan, like the four-door BMW shown above, can carry and conceal a five-hundred pound explosive payload without looking suspicious.

Since cars are an unalterable characteristic of the urban terrain, cars have access to many lucrative targets. If a car can get close to a potential target, the insurgents can also get close to that target with a car bomb. Because car bombs are easy to construct, the insurgents have the ability to employ them on a large scale.

> **In 2005, the insurgents conducted 873 car bomb attacks.**
> **– Brookings Institution Iraq Index**

Chapter 2 – Insurgent Tactics
Car Bombs

Photo by DOD

The car bomb illustrated above (discovered in front of an Iraqi police station) is a relatively unsophisticated example, with several artillery rounds rigged together and placed in the trunk. There were several more artillery rounds hidden under the back seat, connected to the rounds in the trunk. This is a common type of car bomb because artillery rounds are so plentiful across the country and a person with minimal explosives/detonation training can construct one.

Since the payload of this particular car bomb is relatively light (perhaps 400 pounds) and placed in the trunk, the car would appear normal to the casual observer. Only a physical search of the vehicle by security forces would expose it.

This kind of artillery round car bomb has limitations. Several hundred pounds of explosives will cause minimal structural damage and is only effective as an anti-personnel device against relatively "soft" targets.

Chapter 2 – Insurgent Tactics

Aftermath: An Iraqi Police Station

Photo by SPC Katherine Roth

The above picture was taken shortly after a car bomb detonated in the parking lot of an Iraqi police station located in Baghdad, near the Green Zone. More explosives than just half a dozen artillery rounds were used to make this enormous blast.

The insurgents spent a substantial amount of time planning this operation as they were able to infiltrate a vehicle packed with a large explosive payload into a police parking lot. The entire operation probably included police infiltrators, undercover reconnaissance and surveillance agents, a delivery team, and an observation/recording team to conduct post-blast damage assessment and produce propaganda.

The police are a favorite target for insurgents because the police are usually softer targets than military forces. Also, an ineffective police force means insurgent attacks will not be investigated properly, police will be afraid to walk the streets of their communities, and the government's overall human intelligence program will suffer.

Chapter 2 – Insurgent Tactics

Car Bombs – United Nations HQs

(Above are the remains of the United Nations' headquarters building located in Baghdad, August 2003. The building was destroyed by an enormous explosion.)

In order to conduct this successful attack, the insurgents were able to infiltrate the UN building's surrounding security to detonate what must have been an enormous payload – maybe as much as two tons of explosives. In order to cause so much damage the insurgents had to have used some sort of bulk explosives, not a few artillery rounds attached to each other. This was a well planned operation with various insurgent elements working together to pull off the attack.

The insurgents' ability to destroy specific structures and inflict mass casualties forced the United Nations to withdraw from the country. International organizations are favorite insurgent targets as they too are "soft" compared to military targets. A dearth of international organizations serves to isolate the Iraqi government and increases the burden of other security forces committed to staying in the country.

Chapter 2 – Insurgent Tactics

Suicide Car Bombs

Photo by SGT Craig Zentkovich

Above are the remains of a suicide car bomb that killed four American soldiers manning a checkpoint in Najaf, a city located in the southern portion of the country. As can be seen, there is little left of the car after the explosion.

From the insurgent perspective, suicide car bomb attacks are an inherently more secure operation because after the explosion, there is little incriminating physical evidence left behind. If the driver of the car bomb dies in the explosion that is a good thing because that is one less witness alive to incriminate the insurgent organization.

Suicide car bombers have the ability to penetrate security force measures that would stop a less committed person. Suicide car bombers are frequently used to target valuable infrastructure or to inflict mass casualties on soft targets.

> **In 2005, insurgents conducted 411 suicide car bomb attacks.**
> **- Brookings Institution Iraq Index**

Chapter 2 – Insurgent Tactics

Public Service Vehicles

The insurgents use public service vehicles for their operations, like ambulances and police cars, because their semi-official status allows them to pass public scrutiny. Security checkpoints are more likely to let them pass after a cursory look and the average citizen on the roads will pull aside and let them pass.

In the cities, ambulances are a common sight because of the need to treat the victims of widespread daily violence. The insurgents can exploit the ambulance's status and accessibility to the community to facilitate car bomb attacks, to transport guerrilla fighters, and to collect information on the streets.

Ironically, the Red Cross' Baghdad headquarters was destroyed in October 2003 by a suicide car bomber driving an ambulance. In that attack, the ambulance was a perfect way to get past the local security in order to strike the building. Consequently, the Red Cross reduced their presence in the country and eventually pulled out their headquarters staff due to the continuing insecurity. This withdrawal served to isolate coalition forces in the country at the cost of reducing quality medical care available to the public.

Chapter 2 – Insurgent Tactics

Explosive Munitions

There are hundreds of thousands of tons of artillery rounds left over from the former regime that are either stolen, unaccounted for, or unsecured. 152 mm artillery rounds, like those shown above, are the most common and if several of them are daisy chained together, they create a significant blast, capable of destroying armored vehicles. It takes little technical expertise to transform these artillery rounds into IEDs – just detonation cord, a blasting cap, some wire, and a nine-volt battery.

Artillery rounds can be bought on the black market for a reasonable price because they are in such abundance. These artillery rounds are commonly used for making IEDs but can also be used en masse to make a car bomb. Enough explosive ordnance has gone unaccounted for that it is believed the insurgents are able to maintain their current pace of IED and car bomb attacks for the next several years without a need for resupply.

Chapter 2 – Insurgent Tactics

Improvised Explosive Devices (IEDs)

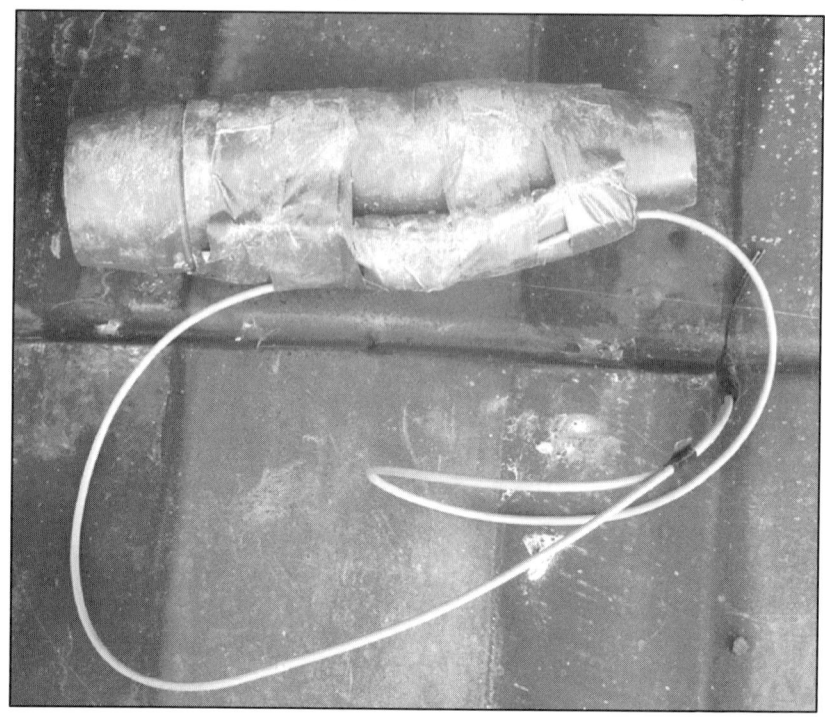

Here is an example of a rudimentary IED – a small, high-explosive artillery round with a pound of plastic explosives taped to the side of it. Detonation cord runs through the plastic explosives and all that is left to do is attach an electric blasting cap to the loop of detonation cord.

This was an above surface IED hidden under debris alongside a road. The intent was to detonate the plastic explosives that would in turn detonate and push the blast of the artillery round across the road and into the target.

An above surface IED this small is only effective against unarmored vehicles like police cars and pick-up trucks or against dismounted soldiers or pedestrians. The simplicity of this device means it can be used on a large scale, requiring minimal technical skills.

Chapter 2 – Insurgent Tactics

Explosively Formed Penetrator

Metal liner.

High explosives.

Metal pipe.

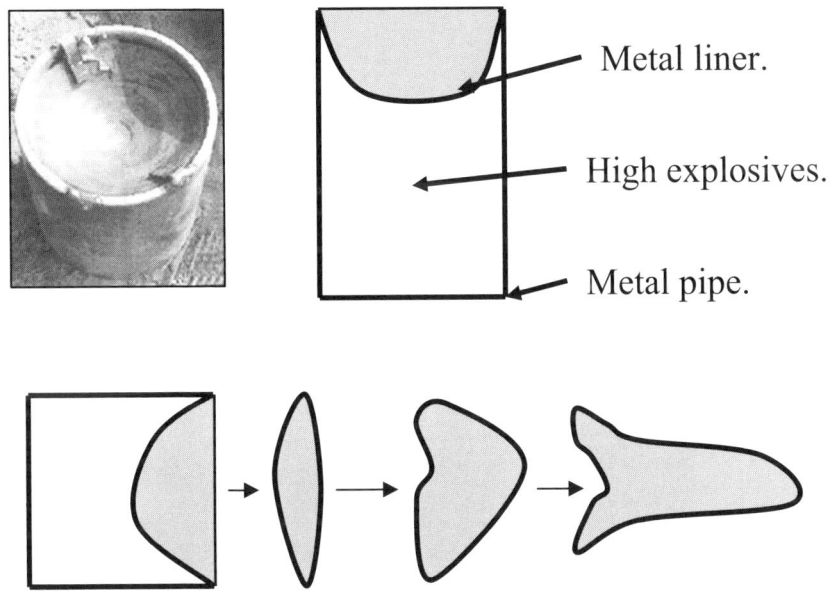

(As the metal liner is shot out of its container, it inverts, becoming a giant bullet-like projectile.)

The explosively formed penetrator (EFP) – also known as an explosively formed projectile – reflects an evolution in IED technology. The above EFP found in Iraq is a common type of EFP used by insurgents in Iraq, believed to be imported into the country from Iran.

The EFP consists of a concave shaped metal liner (often made of copper) which is fitted at the end of a container such as a metal pipe. Behind the bowl-like metal liner is plastic explosives. When the EFP is detonated, the metal liner is launched towards its target. The liner, while en route to its target, becomes inverted, turning into a giant projectile.

An EFP constructed from a metal pipe with a 6" diameter can penetrate over three inches of hardened steel. This penetration power enables the EFP to destroy most armored vehicles such as armored trucks and armored personnel carriers.

Chapter 2 – Insurgent Tactics
Bulk Explosives IEDs

Photo by CPL Bill Putnam

In this picture 600 pounds of high explosives detonate. The insurgents often emplace bulk explosives, such as 500 and 1000 pound aerial bombs, under bridges, in drainage tunnels, and in culverts running under roads. However, in the right terrain and with enough time the insurgents can dig under a road and emplace an IED of any size. While bulk explosives often take the form of aerial bombs, anything can be used like ammonium nitrate or even containers of fuel used in conjunction with high-explosives.

Explosives in this quantity inevitably have fatal effects when detonated under a vehicle. Even armored vehicles like tanks and tracked fighting vehicles will be destroyed. If the vehicle is not destroyed, it may still be blown into the air and flipped over from the blast. The insurgents use these massive IEDs to close off roads to security forces and instill fear in their enemies.

Chapter 2 – Insurgent Tactics

IEDs: Mines

Photo by PFC Elizabeth Erste

The insurgents use factory produced mines as IEDs because they are designed to destroy heavy armored vehicles like tanks. Mines are effective anti-vehicle weapons because they explode where the vehicle's armor is at its weakest – underneath. Even if a mine does not penetrate its target's armor, it may render it immobile by blowing off a wheel or a track.

The insurgents are known to modify these mines by stacking two or three of them on top of each other to multiply their explosive effects. A properly detonated, multiple mine IED will even penetrate the under armor of a main battle tank, destroying it completely.

Also, insurgents regularly attach radio transmitters or cell phones to their mine IEDs so they can detonate them on command. Since many of the landmines the insurgents use are of European design with an almost completely plastic construction, they may be next to impossible to detect when using a traditional mine detector.

Chapter 2 – Insurgent Tactics

IED Ambushes

Photo by DOD

This burned-out supply truck underlines the reality of IED ambushes that are a constant threat in the urban environment. This threat is unavoidable because security forces must travel on the roads to conduct patrols, to set up checkpoints, and transport men and supplies between bases.

Since there are thousands of miles of road in the average city, there are thousands of places for insurgents to set up IED ambushes. Security forces have no hope of identifying and securing the limitless number of places that can be used for an IED ambush.

The IED ambush is an effective guerrilla tactic because it allows the insurgents to set off their IED anonymously and then escape before security forces can react. The IED ambush is the perfect hit and run tactic for the guerrillas who cannot afford to face security forces head on.

The widespread use of IED ambushes by the insurgents have forced security forces to focus a great deal of resources (men, money, equipment) on protecting their road-bound vehicles as opposed to fighting the insurgency.

In 2005, insurgents conducted 10,953 IED attacks.
- Brookings Institution Iraq Index

Chapter 2 – Insurgent Tactics

Weapons Caches

Photo by CPL Mathew Richards

This picture is from a parking garage in Najaf where American soldiers discovered a cache of mortar rounds. The insurgents use caches because they cannot keep incriminating weapons and explosives in their homes or on their property.

The insurgents like to use public property like parks, mosques, abandoned houses, and soccer fields because no individual person owns these locations. Consequently, if security forces locate these caches in a public location, they cannot arrest the owner because the owner is the government.

Chapter 2 – Insurgent Tactics
A Guerrilla Warfare Platform

The insurgents most flexible urban guerrilla warfare platform is the minivan. The minivan is a common vehicle in Iraq's cities and most have curtains covering all of the back windows, concealing one's activities in the back of the vehicle. Vehicles can either be stolen by the insurgents, bought on the black market, or simply purchased at used car lots.

The insurgents use the minivan for a variety of offensive operations like car bomb attacks, drive by shootings, sniper attacks, and transporting insurgents to and from targets that they are conducting raids against.

The minivan is an all around work horse that can be used for a variety of support operations like reconnaissance, surveillance, infiltration, moving long distances, training, and transporting supplies and weapons.

Chapter 2 – Insurgent Tactics

Exploit the Urban Terrain

Photo by FOX News

Every room in every structure is a potential concrete bunker ready to be exploited by either side. In this picture, Marines fire on insurgents using a hotel in Ramadi as a fortress.

The insurgents use the urban terrain as an equalizer to protect themselves from the government's superior firepower. A foot thick wall of concrete will stop any projectile smaller than a .50 round. Insurgents holed up in a concrete structure like this hotel can only be removed by clearing the structure with infantry or high-explosive weapons. An infantry assault will incur casualties and high-explosives cause infrastructure damage. Either way the insurgents score a small victory.

In this example, the insurgents want the government to destroy expensive infrastructure while trying to get to them. Who is going to reimburse the owner of this hotel for the massive damage it is absorbing? The insurgents? The government will end up reimbursing the owner for the repairs.

Chapter 2 – Insurgent Tactics
Infiltration – Day Laborers

photo by SPC ... Rose

The American military and Iraqi security forces depend on local contractors and laborers for a myriad duties – from building mess halls and constructing bunkers to cleaning toilets and removing rubble. Since it is an impossible task to effectively screen the tens of thousands of indigenous workers hired by coalition and government security forces, the insurgents are able to infiltrate government facilities as simple laborers.

The man shown above is being paid to help rebuild a Baghdad police station that was blown up by the insurgents. If this laborer is on the insurgent payroll he will know the entire layout of the new police station and the security measures that are in place. With this information, the insurgents can plan their next attack on the police station before it is even built.

Chapter 2 – Insurgent Tactics

Infiltrate Government Security Forces

The insurgents intentionally instruct their members to join the government's security forces so that they can have eyes inside their enemy's organization.

If the government does not establish a thorough counter-infiltration program, insurgents can sign up off the street and become officially recognized police officers or soldiers.

Most Iraqi security force agencies are infiltrated to some degree, by the insurgents, some massively so.

Photo by LCPL Evan Egan

(In this picture Iraqi men fill out applications for the Ramadi police department. Most screening of potential security force employees is nothing more than verifying information on an application with an existing database.)

Chapter 2 – Insurgent Tactics

Masquerade as Private Security

Photo by LCPL Caleb Smith

Private security forces are abundant in the cities. These security forces have a variety of missions to include checkpoint security, bodyguard duties, convoy protection and security of important infrastructure like oil refineries, airports, and banks. Many of these private security guards wear civilian clothes and openly carry weapons.

The insurgents capitalize on this pattern and themselves dress in civilian clothes and openly carry weapons, acting like security forces running a checkpoint or conducting reconnaissance for a VIP visit. There is nothing to separate the insurgent from the private security force guard except their body language. The general population eyes civilian security personnel with suspicion as they have no idea who they can trust.

Chapter 2 – Insurgent Tactics

Masquerade as Military Personnel

Photo by LCPL Zachery Frank

Because Iraqi military personnel and their families live under the constant threat of insurgent intimidation and assassination, many hide their identities with ski masks, sunglasses, and traditional head scarves.

The insurgents then exploit this fear of exposure and dress like Iraqi security force personnel. Since the security forces wear ski masks as normal procedure, the insurgents too wear ski masks to protect their own identities.

This inability to determine security forces from insurgents creates public mistrust towards the government. The insurgents, dressing like Iraqi military forces, regularly kidnap and execute public employees government workers, and select individuals. The insurgents can get military uniforms as easily as they get police uniforms.

Chapter 2 – Insurgent Tactics

Exploit Cultural Norms

Photo by SGT Jaques-rene Hebert

The insurgents use women for a variety of operations, exploiting cultural norms that place Iraqi women under less scrutiny than men. One advantage for the insurgents is that women can conceal explosives, documents, and money under their traditional garb.

Since women have greater freedom of movement than men do, women can act as couriers and messengers, passing on information from one insurgent organization to another.

Women are an ideal means for conducting reconnaissance and surveillance operations of government security positions because of the deference they receive in Iraqi society.

Chapter 2 – Insurgent Tactics

Surveillance

Photo by DOD

These street vendors in Fallujah are selling their goods at one of the countless roadside markets that dot Iraq's cities. Street vendors like these grow their foods in the countryside but come to the cities to sell their goods. However, selling goods alongside a busy road is an excellent cover for conducting surveillance of government security force activities.

Insurgent surveillance is regularly placed at key intersections, major roadways and highways, and across from the entrances to large military bases so that they can observe military and police activity. Mobile surveillance is often conducted from taxi cabs, people pretending to be out for a walk, and insurgent infiltrators participating in regular security force patrols. There is no way to distinguish a regular street vendor from an insurgent conducting surveillance, making it a widespread problem impossible to eliminate entirely by security forces. Consequently, insurgent surveillance must be assumed, accepted, and its effects minimized.

Chapter 2 – Insurgent Tactics

Reconnaissance

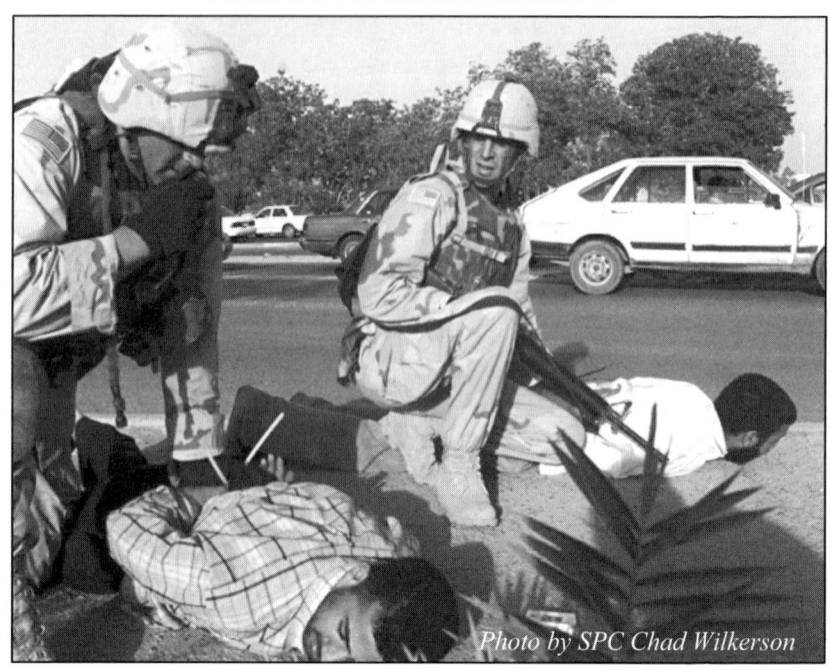

Photo by SPC Chad Wilkerson

The insurgents habitually conduct reconnaissance of future and impending targets using small two and three man teams dressed in civilian clothes, looking like everyone else. These recon teams then report back to their leadership in person or make a cell phone call about what they have seen.

The two Iraqi men above were captured on suspicion of conducting a reconnaissance of security force positions for the insurgents. However, if these two are free of incriminating physical evidence and are not listed in a database as an insurgent, they will have to be released – and the insurgents know this.

As with surveillance operations, reconnaissance operations are impossible to stop because they can be conducted in a manner free of incriminating physical evidence. The insurgents have become experts in reconnaissance, which enables them to successfully plan and conduct their follow on operations.

Chapter 2 – Insurgent Tactics

Fallujah: The Collective Martyr

Photo by CPL Eric Ely

The insurgents are very concerned with fighting and winning the information war as is the government. The graffiti spray painted on the above wall says in Arabic "Long live the Mujahideen of Fallujah".

After security forces retook Fallujah in November 2004, the insurgent spin doctors used Fallujah as a collective martyr. Anyone from Fallujah had instant street credit among the insurgents because of the Fallujans' sacrifices made during the fighting for the city.

The insurgents now demand that their supporters and new recruits show the same commitment as the insurgents in Fallujah did. If this means becoming a martyr for the cause, then so be it. With the proper information spin, even an urban defeat can produce renewed motivation, dedication, and sacrifice from the movement. Some insurgent leaders believe that the events in Fallujah have made their movement even stronger.

Chapter 2 – Insurgent Tactics

The Mosque System: Strategic Location

Photo by DOD

This aerial view of a mosque in Fallujah reveals the mosque's strategic positioning in every Iraqi city. The mosque is a centrally located facility, ensuring that the surrounding residential areas and people have easy access to it.

Consequently, the mosque system is key terrain and its successful exploitation either by insurgents or the government can have strategic effects. An insurgent controlled mosque allows the insurgents to penetrate the local community and create their own grass-roots political movement.

Furthermore, an insurgent controlled mosque is an enormous resource for the insurgents because it provides a direct link from the insurgents to the surrounding community. An insurgent controlled mosque can spread the insurgent message directly to the people in a variety of ways to include sermons by the imam or through the mosque's loudspeakers. Since a mosque's imam exerts much influence in the community, the insurgents go to extra lengths to recruit and then protect imams who support the movement.

Chapter 2 – Insurgent Tactics

Mosque Exploitation

Photo by LCPL Kaemmerer

Photo by DOD

Above, soldiers burn insurgent propaganda confiscated inside a mosque. Insurgents regularly use the mosques as a platform to push their political message to the people.

To the left, American soldiers display a weapons cache found inside an insurgent mosque in Baghdad. The insurgents use the mosques (which are community property) to hide weapons, explosives, and other equipment.

Chapter 2 – Insurgent Tactics
Civil Disturbances

Photo by LCPL ... han Heusden

The high concentration of people living in the cities allows the insurgents to exploit perceived and real injustices committed by government security forces against the public.

Since the insurgents have access to large groups of people living in geographically small areas, the insurgents can organize and excite emotional crowds to riot, causing infrastructure damage by burning cars and looting structures.

The above angry crowd in Najaf is demonstrating against the killing of a local citizen by American forces. The green headbands indicate that they are likely members of Moqtada Al-Sadr's Mahdi Militia.

Because Iraqis are emotional people, influential tribal and religious leaders can manipulate the public to support the insurgent agenda. Professional agitators can try and provoke security forces to fire on a disruptive crowd in order to manufacture a massacre that can then be trumpeted in the insurgent's information campaign.

Chapter 2 – Insurgent Tactics

Organized Crime

Photo by SPC Ryan Smith

In this picture American soldiers arrest a local Iraqi accused of carjacking. The insurgents have a natural affiliation with organized crime because both are committing crimes against the state.

The more effort the government spends fighting crime, the more breathing space the insurgents have. The reverse is also true for a police force focused on fighting a guerrilla war has less time to fight the common criminal.

The insurgents benefit in a variety of ways by associating with organized crime. Criminal gangs can provide the insurgents with information, hostages, stolen cars, money and a variety of other services like strong-arming and intimidating mutual enemies.

Both organized criminal gangs and the insurgents use similar methods and tactics to foil the police and continue their illegal activities. In many ways an insurgent organization mirrors the organizational structure and methodology of a criminal gang.

Chapter 2 – Insurgent Tactics

Insurgent Medical Care

(Above is a picture of Saddam Hospital located in the southern city of Nassiriya where American soldier Jessica Lynch was rescued from in April 2003.)

Since the insurgents dress, look, and act like the surrounding populace they can receive medical treatment from government run hospitals. If an insurgent is wounded in combat, they can approach security forces and tell them that they were shot by the insurgents for helping the government.

As long as the insurgent has no incriminating physical evidence on their person and is not in a security force database, the insurgent will get quality medical treatment and then be released. Wounded insurgents can also go to insurgent run clinics and get treated with no questions asked.

Chapter 2 – Insurgent Tactics

Kidnappings

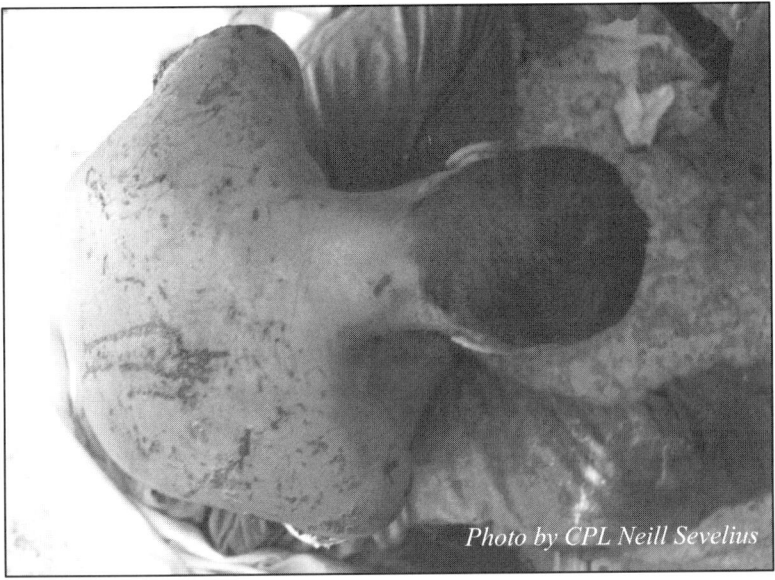

Photo by CPL Neill Sevelius

The Iraqi man in this photograph bears the scars after prolonged torture and interrogation by insurgent captors. American forces found him in a torture chamber, after having been hung upside down by his feet, electrocuted, and beaten.

In the cities, the insurgents have ready access to all kinds of valuable people like foreign aid workers, journalists, businessmen, government officials, and coalition soldiers.

The insurgents take hostages for many reasons like extracting information, extorting money, to apply political leverage against a country, to sell the victims to another insurgent group for a profit, and to instill terror in a specific segment of society.

By December 2005, 30 Iraqis a day were kidnapped across the country. Through March 2006, 280 foreigners had been kidnapped in Iraq.
– Brookings Institution Iraq Index

Chapter 2 – Insurgent Tactics

The Face of Urban Terror

Abu Musab Al-Zarqawi DEAD

Photo by DOD

For three years, Abu Musab Al-Zarqawi was the face of urban terror in Iraq until his death in early June 2006. With a $25 million bounty on his head and his face plastered all across the nation on reward posters, Zarqawi was the most wanted man in Iraq.

Zarqawi perfected the techniques of urban terror – car bombs, suicide bombers, hostage taking, mass murder, and high-profile internet executions – to exact a bloody toll.

Cities, with their high concentrations of people in public places (markets, buses, funerals, religious services, crowded streets), offer an endless array of targets for terrorists like Zarqawi who commit mass murder to destabilize the country and incite civil war.

> **By March 2006, 5,683 Iraqis were killed and another 11,141 wounded in multiple fatality bombings.**
> **– Brookings Institution Iraq Index**

Chapter 2 – Insurgent Tactics

The International Jihad

Picture from a Bin Laden video

Picture from a Zarqawi video

Usama Bin Laden has a vested interest in seeing the Iraqi insurgency succeed and coalition efforts fail. To this end, Bin Laden supports international Islamic jihadists like the Jordanian-born Zarqawi. In this relationship, terrorist and insurgent groups operating within Iraq benefit from Al-Qaeda's financial support while, Al-Qaeda reaps the benefits of association with groups fighting the Great Satan – The United States.

Chapter 2 – Insurgent Tactics

The Human Toll

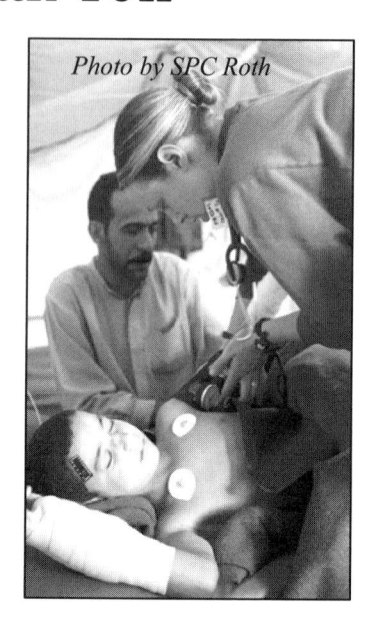

Photo by DOD

Photo by SPC Roth

An estimated 33,489 to 37,589 Iraqis died from March 2003 to March 2006.
– www.iraqbodycount.org

95 out of every 100,000 Iraqis are murdered annually in Baghdad (a total of 24,700 if this rate is applied country wide.)
- Brookings Institution Iraq Index

2,390 American soldiers have been killed and 17,500 wounded from March 2003 until April 2006.

4,219 Iraqi soldiers and police officers have been killed. (In 2005 alone 1,497 Iraqi Police Officers were killed and 3,256 wounded.)
- Brookings Institution Iraq Index

Chapter 3
People: The Key Terrain

Photo by SGT April Johnson

Chapter 3 – People: The Key Terrain

Interpreters: A Vital Link

Photo by CPL Mike Escobar

Speaking the local language is critical for understanding and interacting with the populace. A guerrilla war cannot be won unless security forces are able to communicate with the people. In this arena, the insurgents have a huge advantage over foreign military forces that cannot talk to the people on the streets. The insurgents do not need interpreters and benefit greatly by understanding the nuances of their own language.

Therefore, local interpreters are an important linchpin between foreign security forces and the people and the insurgents understand this. The insurgents seek to sever the link between American forces and the Iraqi people by getting rid of the go between – the interpreter. Notice how the above interpreter in Fallujah has his face completely covered so that the woman the interpreter is speaking to cannot recognize him. The need for interpreters to hide their identity is due to the fact that the insurgents have made it a policy to assassinate any interpreter working for the security forces.

Chapter 3 – People: The Key Terrain

The Importance of Media

Photo by SSG Angelique Perez

Journalists and reporters from all over the world, working for a variety of newspapers, television and radio stations, flock to the cities to get the latest scoop. Much of the Iraq war is covered by reporters who live in the Green Zone in central Baghdad.

Both the insurgents and security forces are acutely aware of the media's presence and use it to their advantage. Insurgents often find a sympathetic ear from Middle Eastern media giants like Al-Jazeera and Al-Arabiyah while American forces have ready coverage from CNN and Fox News.

International media coverage is a critical component of the war because what is reported affects the world's perceptions of and attitudes towards the ongoing war.

> **In October 2005 Iraq had 44 commercial TV stations, 72 commercial radio stations, and over 100 independent newspapers and magazines.**
> **- Brookings Institution Iraq Index**

Chapter 3 – People: The Key Terrain

The Information War

A Marine spray paints pro-government graffiti on a wall in Ramadi. Graffiti is one way for security forces to get their message out to the people. If the message is spray painted in a busy market area, several thousand people a day will see it.

Photo by CPL Tom Sloan

These soldiers are placing their posters over anti-government graffiti spray painted on the wall by insurgents. The insurgents spread their message during the night and the government spreads their message during the day, creating a continuous, information warfare cycle.

Photo by SGT Jeremiah Johnson

A soldier distributes pro-government flyers to people in Fallujah. The local populace must be careful since receiving such flyers in an insurgent run neighborhood may be a death sentence. The insurgents also distribute CDs and DVDs on street corners and through the black market. The insurgents' distribution machine should not be underestimated.

Photo by PFC Michael Cardin

Chapter 3 – People: The Key Terrain

Influencing the Population

Photo by SSG Quinton Russ

One of the government's best means to positively influence the population is through civil affairs activities. A government that improves the health and overall quality of life of its citizens will enjoy increased legitimacy. Plus, it is difficult for the insurgents to say that curing an ailing little girl is a bad thing.

Medical civil affairs projects may not uniformly improve the nation's health, but it shows that the government cares by putting a human touch on a counterinsurgency campaign that is often bloody. Importantly, people who have their health improved by security forces are more willing to provide the government with tips about what is going on in their neighborhoods.

The insurgents too understand the importance of helping the people and the insurgents conduct their own civic action projects, funding public health clinics, giving grants to needy people, and essentially acting as a shadow government.

Chapter 3 – People: The Key Terrain

Good Relations Are Crucial

Photo by SGT April Johnson

This picture is a good example of why vehicle patrols, static checkpoints, and other population control measures will never defeat an urban insurgency. Only intimate, positive interaction with the population through community policing will yield results.

The dismounted foot patrol is an urban insurgency's greatest threat. Security forces that take time to talk with the people and show them goodwill develop their own grass-roots human intelligence network. This human intelligence network will consist of housewives, kids, and other citizens who want to bring security to their neighborhoods and stability to their lives.

Civic action is one way to develop good relations with the people, but so is understanding that every soldier is an ambassador of goodwill. Since Iraqis are intensely personal people, true friendship often trumps political and military agendas.

Chapter 3 – People: The Key Terrain

Intelligence Collection

(Security forces question a local man during a nighttime operation.)

Human intelligence is the most important and effective means of collecting information in an urban environment. Only humans can go inside apartment buildings, take a ride in a taxicab, walk inside basements, and pray inside mosques. Only humans can determine someone's intentions by reading their eyes or studying their body language – electronics and machines cannot.

Successful intelligence gathering through human assets requires personnel trained in human intelligence methodology, interpreters, money, and close ties with the community. Anything short of a well supported, comprehensive human intelligence network falls flat. An urban guerrilla war is human intelligence intensive, on both sides, and is the key to a successful (or failed) military campaign.

In December 2005, security forces received 3,840 actionable tips from the populace.
- Brookings Institution Iraq Index

Chapter 3 – People: The Key Terrain

Population Control - Checkpoints

Photo by SPC Darryl Magby

The checkpoint is a ubiquitous feature of any city and is part of the urban terrain. The checkpoint is a population control measure that attempts to screen out insurgents from the rest of the population. At checkpoints cars are searched for explosives, weapons, and other contraband while passengers and drivers are questioned.

Insurgents benefit from checkpoints because checkpoints inconvenience the population when the people are searched. Checkpoints also back up traffic and make people late for work. More importantly, checkpoint guards are convenient targets for insurgent sniper, drive-by, and car bomb attacks.

The insurgents study the government's checkpoint system and learn to bypass them or pass through them undetected. Since a city has thousands of roads and alternate routes, insurgents are usually able to bypass a particular checkpoint and still reach their destination. It would take a massive effort by security forces to create a truly air tight system of checkpoints.

Chapter 3 – People: The Key Terrain

Population Control - Dogs

Photo by SSG Monica Garreaux

Dogs are an extremely useful tool in the urban environment because they can smell things that humans cannot detect using other means. Dogs are ideal for sniffing out high-explosives and even gun powder residue. Because the insurgents conceal explosives in such ingenious ways that only a dog can detect them, dog teams are vital.

Most conventional security forces neglect training and employing working dogs at the level required to significantly impact operations. While dogs have limited benefit on the conventional battlefield, they are a key piece for an urban guerrilla battlefield. A few dog teams contribute very little, but employed on a large scale with hundreds employed at checkpoints, during raids, and on foot patrols they become a counterinsurgency force multiplier.

Chapter 3 – People: The Key Terrain

Population Control – ID Screening

Photo by LCPL Paul Robbins Jr

Local law enforcement are the right personnel to screen insurgents from the population. Indigenous security personnel know the language, know the people, and know the customs. They are better able to identify suspicious behavior or an attempted cover story that does not add up. Local law enforcement can also determine where a citizen is from by their accent – if they are from the north of the country, from the south or from another country like Syria or Iran. A foreigner will never pick up on these nuances.

Because checkpoints thoroughly scrutinize identification cards and passports, the insurgents have become experts in creating and using fake documents. A robust black market industry has blossomed in Iraq and getting fake ID cards and passports is easy and inexpensive. Even if a checkpoint does not successfully screen an insurgent, at least security personnel can ensure they are not carrying any weapons or explosives.

Chapter 3 – People: The Key Terrain

Population Control – Local Customs

Photo by DOD

In Iraq it is socially taboo for men to physically search women and children. Consequently, women and children generally receive less scrutiny at government checkpoints. The insurgents exploit this social norm and exploit breaches of government security by using women and children to serve as couriers to transport documents, information, and even explosives.

It is not surprising that the insurgents are increasingly using women as part of their organization. Women also make the best suicide bombers because local customs increase their ability to penetrate government security measures as compared to their male counterparts.

Security savvy organizations employ female guards at checkpoints for the sole purpose of more thoroughly scrutinizing women and children. Female inspectors are an astute move, respecting local customs and norms but still maintaining security.

Chapter 3 – People: The Key Terrain

Population Control – Establishing Identities

Photo by SSG John Knauth

Because quality fake IDs and passports are so common, more technical means of identification are required to detect potential insurgents. Retinal scans are one sure method to establish a person's identity. Retinal scanners are useful not only at checkpoints but at detention centers and even during raids when positive identification of a person is required.

There is no way to fake a retinal scan and the scan will at least ensure that a potential insurgent uses the same set of identification papers even if that original set itself is fake. Other means of identification more reliable than checking documents include fingerprinting and DNA testing.

Much of a government's counterinsurgency campaign is an identity war where the government tries to get accountability of each and every member of the population. In response, the insurgents do everything in their power to remain anonymous.

Chapter 3 – People: The Key Terrain
Population Control – Databases

Photo by CPL Mike Escobar

This Marine, as he inputs information into his laptop, is part of a larger effort to control the population by establishing an accurate, widespread database of people living in a specific area. An insurgent can only be identified at a checkpoint or during a raid if he is identified as one in a database. A database is a living document, takes years to build, and must be updated daily.

The challenge for government security forces is developing a national data base that all security officers have immediate access to. A person may be identified as an insurgent in the Fallujah database, but if police at a checkpoint in Baghdad do not have this information at their fingertips, this same insurgent came operate easily in the capital.

Facilities housing population databases are lucrative targets for the insurgents. One reason why police facilities are frequent targets of insurgent attacks is because local law enforcement agencies have the most intimate knowledge of the surrounding communities.

Chapter 3 – People: The Key Terrain

Population Control – Forced Evacuations

Photo by DOD

(A deserted street somewhere in Fallujah.)

One method government forces use to separate the insurgents from the population is by evacuating an entire city. Before American forces retook Fallujah in November 2004, the entire city's population of 300,000 people were forced to leave. The government's stance was if all the innocent people left the city, only the insurgents would remain.

However, this mass evacuation is not a perfect solution. This same methodology gives the insurgents advance warning of an impending operation, allowing them to leave along with the population. Additionally, not everyone will voluntarily leave the city. Some people will remain behind with their homes.

Separating the people from the insurgents does allow security forces to conduct high-intensity operations with less chance of killing innocent people. It should be remembered that these same security forces are more likely to now resort to infrastructure damage once this separation of the insurgents and the people is achieved.

Chapter 3 – People: The Key Terrain

Raids: Removing the Insurgents from the Populace

Photo by SGT Jeremiah Johnson

(Security forces prepare to enter and clear a structure during a nighttime raid.)

The nighttime raid is a fundamental characteristic of urban guerrilla warfare as government security forces attempt to hunt down and capture suspected insurgents. As such, the raid is not necessarily intended to defeat an insurgent force as much as it is another effort to remove the insurgents from the population.

Security forces find themselves conducting a never ending series of raids as they try and find the estimated 20,000 insurgents living among Iraq's 26 million people.

Insurgents attempt to limit the effectiveness of government raids by infiltrating the organizations conducting the raids, by living a life on the run, and enforcing a "clean" life free of incriminating physical evidence.

Chapter 3 – People: The Key Terrain

Household Searches

(Soldiers conduct a search of a house during a raid.)

A common task for every security force hunting for insurgents is conducting household searches. A household search may take place after a raid or security forces may simply knock on the door of a local residence and conduct a random search.

The household search is part of the battle for incriminating evidence where security forces attempt to link suspected insurgents with crimes against the state with the physical evidence found in their homes. In response, insurgents develop procedures to avoid being caught with incriminating evidence like using weapons caches located in public places like mosques, cars, and abandoned homes.

Regardless of the outcome, if security forces search the wrong house, insult an innocent family, and steal or destroy their property, they may create new insurgents instead of hurting the guerrilla movement.

Chapter 3 – People: The Key Terrain

Detainees

Photo by DOD

If security forces do capture a known insurgent in a raid, do find incriminating physical evidence during a household search, or do arrest an insurgent at a checkpoint, these people are detained and enter the legal system. From this initial contact with security forces, the insurgents are trained to create doubt in the mind of their captors by professing their innocence. Insurgents are also trained to separate themselves from incriminating physical evidence, never keeping anything on their person.

If taken into custody, detainees are transferred to a detention facility for further questioning and processing. Retinal scans and fingerprinting will ensue to positively identify the detainee and enter him into a database. After processing, the insurgent loses his anonymity and becomes a known element, making future insurgent activities more difficult for them. For the insurgents, the war continues, even after capture. From this point on the insurgent uses their arrest to further the cause and get back onto the streets. In the prisons, insurgent groups become a more organized, better trained, stronger movement.

> **Since August 2005 2,800 detainees have been released from the detention system.**
> **- Brookings Institution Iraq Index**

Chapter 3 – People: The Key Terrain

Detainees: A Growth Industry

Photo by Michael Larson

In January 2006 American forces were holding an estimated 14,000 detainees in various prisons like the infamous center at Abu Ghraib.

In December 2005, Iraqi authorities were holding an additional 12,000 detainees in Iraqi controlled prisons.

330 foreign fighters had also been detained.
- Brookings Institution Iraq Index

Chapter 3 – People: The Key Terrain
Political Campaigning & Elections

(An American soldier on patrol somewhere in Baghdad, with election posters plastered on a wall behind him.)

Ultimately, influencing the population through information operations and removing the insurgents from the populace through raids and other population control measures will enhance a process where the people elect a legitimate government.

(A man shows his ink-stained finger – proof that he has voted.)

However, as of April 2006, organized, democratic elections had yet to weaken the insurgency or bring security and stability to the country.

A January 2006 poll showed that 47% of Iraqis approved of insurgent attacks on U.S. forces in Iraq and only 6% of Sunnis thought the country was headed in the right direction.
- Brookings Institution Iraq Index

Chapter 3 – People: The Key Terrain

Polling Stations

Photo by LCPL Shane Keller

Government security forces have an enormous burden in the cities on voting day. Millions of people have to be protected as do thousands of different polling stations.

Lines of voters are tempting targets for insurgents bent on committing acts of urban terror. Car bombs and individual suicide bombers are the insurgents' preferred methods for killing large numbers of people queued up in long lines.

In order to provide effective security on voting day for the entire voting process, every security asset is required. But the effort is justified as there is no more important task than enabling the people to choose their own futures – especially if open elections are part of the government's counterinsurgency plan.

In a January 2006 poll, 87% of Iraqis approved the government endorsing a timeline for a U.S. withdrawal.
- Brookings Institution Iraq Index

Chapter 4
Weapons of War

Photo by SSG James Christopher III

Chapter 4 – Weapons of War

The M16 and M4

Photo by DOD

The M16 is a rural-centric weapon that is inappropriate for urban warfare. Long rifles like the M16 cannot be concealed or maneuvered in tight spaces like vehicles and enclosed buildings. Additionally, the 5.56 mm round lacks the ability to penetrate the mediums commonly found in an urban environment like body armor, vehicles, doors, windows, and walls.

Photo by SSG Rebekah-mae Bruns

The M4 with 40 mm grenade launcher is a much better urban weapon system. The M4 is a shorter weapon, allowing it to be better maneuvered in the confines of an urban environment. However, the telescoping stock is still too long and should be replaced with a folding stock. The grenade launcher allows the firer to destroy vehicles and penetrate mediums like walls and doors, so it can be used as a breaching tool when required.

Chapter 4 – Weapons of War

The M240 and M249

Photo by SSG Klause Baesu

The M240 is a reliable 7.62 mm general purpose machine gun used in a variety of roles, either mounted on vehicles or carried by dismounted infantry. The 7.62 mm round has penetration power, enabling it to penetrate the mediums commonly found in an urban environment like body armor, vehicles, doors, and walls. Like all full-sized machine guns, the M240 is too large and heavy to be maneuvered deftly in an urban environment.

Photo by LCPL Shane Keller

The M249 light machine-gun shown here has been modified for the urban environment with a shortened barrel, an adjustable stock, and an attached ammo bag with a 100 round belt. The modified M249 is maneuverable in confined spaces - a good balance of compact size and sustained firepower, albeit of the 5.56 mm size. It is good for a variety of roles to include CQB, which is rare for a belt fed machine gun.

Chapter 4 – Weapons of War

The Shotgun and Pistol

Photo by SSG Joseph Roberts

The 12-gauge shotgun is a must for the urban environment because of its importance as a breaching tool. The shotgun will not overpenetrate urban mediums so the risk of collateral damage is reduced. However, full stock shotguns are inappropriate for the urban environment. Pistol grip shotguns with no buttstocks are required so the shotgun can be carried in vehicles and backpacks and maneuvered in tight spaces.

Photo by SPC Sean Kimmons

Every soldier requires a pistol in an urban environment for a variety of reasons. A pistol is a must for a back-up weapon, can be fired from inside vehicles, and many urban warfare problems (like searching cars, closets, under beds) demand the use of a pistol. However, the 9 mm Berretta pistol shown above lacks the ability to penetrate urban mediums – a .45 pistol is demanded in the cities.

Chapter 4 – Weapons of War

The MK19 and Claymore Mine

Photo by SSG Klaus Baesu

The MK19 is an 40 mm automatic grenade launcher. It can penetrate concrete walls and other urban mediums. It is useful for destroying cars or reducing enemy fighting positions. It is a high-intensity urban warfare tool, not a precision weapon.

Photo by SPC Gul Alisan

Because the claymore mine is an area weapon, it has limited utility in an urban guerrilla war. However, it can destroy unarmored vehicles and is effective against enemy infantry. It is also useful when employed in a defensive manner inside buildings to detonate down a hallway, down a stairwell, or into a room that has just been cleared by an element of dismounted enemy personnel.

Chapter 4 – Weapons of War

The Marksman

Photo by Shane Cuomo

The above marksman is using an M-21/25 sniper rifle that fires the standard NATO 7.62 x 51 mm round. This rifle allows the shooter to penetrate body armor, cars, windows, and doors.

A scoped rifle allows the marksman to identify specific targets, making the marksman a precision weapon system. The ability for precision allows the marksman to minimize collateral damage – he only hits what he intends to hit.

Because of the high-power scope and accurate rifle, the above soldier can engage relatively small targets at longer ranges than a soldier with an M16 or M4. The marksman can reliably hit an enemy in the head at 200 meters and in the body at 600 meters.

Chapter 4 – Weapons of War
AT4 and .50 Heavy Machine Gun

Photo by SPC Sean Kimmons

The AT4 is a good high-intensity urban warfare weapon. The AT4 will blow up cars, armored vehicles, and small buildings. The AT4 can be used to breach a structure or take out a pillbox. However, it has very limited use as an urban counterinsurgency tool due to its high-intensity nature.

Photo by Mike Buytas

The Browning .50 heavy machine-gun (HMG) has good penetration, capable of destroying cars and lightly armored vehicles. It will shoot through concrete walls and reduce enemy fighting positions to rubble. The .50 is a high-intensity warfare weapon with limited use in an urban guerrilla war because of its overpenetration. The .50 HMG can be configured in a number of ways, either fired by a gunner or remotely (as shown above). The HMG can also be mounted on a variety of platforms such as an armored truck, armored personnel carrier, or a tank.

Chapter 4 – Weapons of War
Indirect Fire Systems

Photo by DOD

Photo by SGT Ashley Rice

Photo by SSG James Christopher III

Indirect fire systems are intended for high-intensity warfare. Mortars are useful in high-intensity urban warfare because the can fire over buildings and other urban obstacles, hitting enemy personnel on the other side of them. Artillery can be used to destroy single buildings with guided munitions. Both can fire illumination rounds to light up the urban battlefield. Rocket launchers can destroy an entire city block when required. All of these systems can fire air burst projectiles, which is useful for cleaning the roofs of enemy personnel.

Because indirect fire systems are area weapons by nature, they have limited utility in an urban guerrilla war where the goal is to minimize infrastructure damage and civilian casualties. Urban guerrillas intentionally conduct their own indirect fire attacks in the cities to provoke an artillery barrage from security forces. If security forces do respond with their own artillery, the insurgents will use the resulting collateral infrastructure damage and civilian casualties to fuel their information war.

Chapter 4 – Weapons of War

The Main Battle Tank (MBT)

Photo by SPC Danielle Howard

The Abrams MBT is an excellent high-intensity urban warfare tool. It gives its occupants 360 degree protection and has a variety of weapons systems to include a .50 heavy machine gun, two 7.62 mm machine guns and a 120 mm main gun. The MBT can be used to sweep roads for IEDs, establish checkpoints, and support infantry operations.

Photo by LCPL James Vooris

However, the MBT is ideal only as a high-intensity warfare tool. Its main gun will destroy an entire building – hardly appropriate when a surgical, restrained response is required. Also, the MBT's occupants have poor situational awareness and cannot interact with the populace. The MBT is seen by the populace as a symbol of escalating militarization – not normalcy and stability.

Chapter 4 – Weapons of War

Wheeled Armored Fighting Vehicles

Photo by SGT Jeremiah Johnson

The Stryker armored fighting vehicle shown above gives its occupants 360 degree protection and has a remotely fired .50 machine gun or a 40 mm automatic grenade launcher. This remotely fired weapon system allows the vehicle's occupants to remain under cover while engaging enemy threats. As with all enclosed armored vehicles, the occupants have reduced situational awareness, cannot interact with the populace, and represent an increased level of militarization.

Additionally, the Stryker is better suited for open terrain because it is so large and ungainly. The Stryker can only travel on main roads that are wide and open. Turning this vehicle around creates problems on anything less than double or triple lane roads. Vehicles this large cannot travel down side roads, alleys or other restricted roadways. More maneuverable vehicles are better suited for the confines of the urban environment.

However, since the Stryker is a wheeled vehicle (eight wheels, all wheel drive), it is quieter than a tracked vehicle and its tires cause less infrastructure damage to a city's streets, sidewalks, and medians than does a tank or similarly tracked vehicle.

Chapter 4 – Weapons of War

Tracked Armored Fighting Vehicles

Photo by SPC Mary Rose

The Bradley armored fighting vehicle shown above gives its occupants 360 degree protection and has a variety of weapon systems to include TOW antitank missiles, a 7.62 mm machine gun and a 25 mm main gun. The 25 mm gun has excellent penetration capability and is very accurate due to the Bradley's sophisticated fire control system. The Bradley has the ability to transport a team of infantryman inside, providing them with 360 degree protection until they exit the vehicle.

The Bradley, as a tracked vehicle, can pivot turn in place, allowing it to maneuver in restricted areas. All tracked vehicles cause some level of infrastructure damage, tearing up paved roads, curbs, and sidewalks. This infrastructure damage in turn costs money to repair, is an inconvenience to the local population, and undermines the government's legitimacy.

As with all enclosed armored vehicles, the occupants have reduced situational awareness, cannot interact with the populace, and represent an increased level of militarization. The Bradley is a loud vehicle and can be heard several kilometers away as it approaches its target area. The Bradley is a good high-intensity warfare vehicle, but less appropriate for a guerrilla war.

Chapter 4 – Weapons of War

The Armored Truck

Photo by Mathew Wester

The High Mobility Multipurpose Wheeled Vehicle (HMMWV or Humvee) is an armored truck with two-inch thick bullet resistant windows and ¼ inch steel armor that defeats 7.62 mm rounds. This vehicle is four-wheel drive with high ground clearance, allowing it to drive over curbs, sidewalks, and road medians. Because it is a wheeled vehicle like the Stryker, the Humvee causes less infrastructure damage and is quieter than tracked vehicles. The Humvee is an excellent general purpose workhorse.

Depending on the vehicle model, the Humvee may have a turret gunner or a remotely fired weapon system. As with all enclosed armored vehicles, the occupants have reduced situational awareness, cannot interact with the populace, and represent an increased level of militarization. However, the Humvee has a less militarized signature as compared to a tank or armored fighting vehicle. Also, the crew has a choice to emplace or remove a turret mounted machine-gun, depending on the image they want to present.

Chapter 4 – Weapons of War

RG-31 Cougar

Photo By DOD

The Cougar, a South African design, is an excellent all around urban warfare vehicle because its "V" shaped hull is designed to deflect the upwards blast from IEDs. The vehicle shown above was damaged in an IED blast but its occupants all survived, suffering nothing more than concussions and other minor injuries.

Another excellent feature of the Cougar are the gun ports located along its bulletproof windows. This allows the soldiers inside it to shoot outside without opening a window (unlike the Humvee). Also, the number of windows allows for 360 degree observation (unlike the Humvee) resulting in improved situational awareness.

Because the Cougar has tires it is quiet and causes no infrastructure damage. However, this four-wheel drive vehicle can still drive over curbs, mediums, debris and other urban obstacles. The Cougar can comfortably carry five people inside and newer models are being fitted with remote controlled weapons systems allowing the crew to engage targets while remaining protected inside the vehicle.

Chapter 4 – Weapons of War

Police Cars

Photo by CPL Heido Loredo

Police cars are useful for conducting routine security patrols, for showing a security force presence, and giving the people a sense of normalcy and stability. Local law enforcement are the first line of defense against an insurgency since they interact with the surrounding communities on a daily basis.

However, police cars are only appropriate for security situations where there is a relatively low risk to uniformed police officers. Since police cars are identifiable as government security force vehicles, they are lucrative targets for the insurgents.

The insurgents intentionally target police vehicles because they are unarmored, soft targets. The insurgents want to force the police off the streets so that the government has to respond with armored, military vehicles, creating an image of instability and insecurity. The best answer to this problem is for local police forces to purchase and employ civilian manufactured armored vehicles so they are no longer "soft" targets.

Chapter 4 – Weapons of War

The Modified Truck

Photo by CPL Mike Escobar

Iraqi security forces frequently use civilian pick-up trucks for conducting security patrols. These trucks are good for rapidly transporting large numbers of soldiers (up to a dozen in a truck) from one place to another and are often used for quick reaction duties. These trucks have all the benefits of normal, wheeled civilian vehicles: they are quiet, cause no infrastructure damage, and from a distance look like a non-military vehicle. Because these are slightly modified civilian vehicles, they project a less forceful image.

Security forces often modify these trucks, installing a machine gun mount in the bed of the truck, turning them into a support by fire vehicle. However, once a machine-gun is mounted on the vehicle and a uniformed gunner stands up in the bed of the truck, they are quickly identified as a security force vehicle. These vehicles are vulnerable to IED, car bomb, and small arms ambushes because they are unarmored. As with the standard police car, these vehicles are inappropriate for a high-intensity warfare environment.

Chapter 4 – Weapons of War

The Nonstandard Tactical Vehicle (NSTV)

The NSTV is nothing more than a civilian vehicle that is a more appropriate vehicle used to reduce a military presence, blend in with the urban environment, and show an increased level of normalcy. NSTVs can be armored to defeat up to 7.62 mm rounds and SUV models, like the one shown above, can still drive over curbs, medians, and other urban obstacles. The NSTV is best used with personnel in civilian clothes who also blend in with the urban environment. The NSTV can be used in conjunction with specially trained personnel for a variety of operations such as reconnaissance, surveillance, sniper, and infiltration missions as well as high-risk raids.

The NSTV must be kept in perspective – it is not a high-intensity warfare tool, it is a counterinsurgency platform. While a tank or armored fighting vehicle may survive an IED, car bomb, or RPG attack – the NSTV probably will not. The NSTV can only work through deception by blending in with the urban environment. Once the NSTV no longer blends in, it is exposed and becomes a relatively soft target (as compared to an armored truck with a heavy machine gun).

Chapter 5
Aircraft in the Cities

Photo By CPL Mathew Apprendi

Chapter 5 – Aircraft in the Cities

AC-130 Gunship

Photo by DOD

The AC-130 gunship is an excellent high-intensity urban warfare tool. Because the AC-130 has down-looking, infrared, gun cameras the aircraft's crew has an excellent, bird's eye view of the urban terrain during the day or at night.

The AC-130 also has several different weapons systems (depending on the gunship model) it can use to engage targets. These systems include a 25 mm gatling gun, a 40 mm cannon, and a 105 mm howitzer. By having a choice of weapons, the gun crew can select a specific weapons system to engage a specific target in order to limit the amount of collateral damage inflicted.

The AC-130 can be used not only for fire support, but as eyes in the sky, providing situational awareness for soldiers conducting operations in the streets below. Because the AC-130 can fly in a relatively slow, tight racetrack, its crew can keep both its weapons and observation systems focused on the urban terrain below, enabling the crew to retain their situational awareness and stay locked on a target.

Chapter 5 – Aircraft in the Cities

High Performance Aircraft

Photo by CPL Paul Leicht

Attack aircraft like the F/A-18D Hornet have the ability to provide accurate strikes in urban areas with guided weapons like the JDAM systems carried on the Hornet illustrated above. The JDAM (or Joint Direct Attack Munition) is an add on kit that modifies an existing "dumb" bomb into a "smart" GPS guided bomb. These GPS guided JDAMs are able to strike individual targets within ten meters of the intended point of impact.

However, any guided aerial bomb strike is only applicable for destroying a specific structure and therefore has limited application in an urban guerrilla war. For example, if an insurgent fires on a security force patrol from the third floor of a five-story apartment building an aerial bomb is useless because dropping a precision munition will still destroy the entire building and the scores of innocent people in it. All aerial dropped munitions suffer from this same limitation.

High-performance aircraft are more useful than just dropping guided munitions since many can be fitted with surveillance packages, allowing them to record real-time imagery of the streets below with top-down looking, all weather cameras.

Chapter 5 – Aircraft in the Cities

AH-64 Apache

Photo by TSGT Andy Dunaway

Like all aircraft, attack helicopters benefit from a bird's eye perspective of the urban terrain. Helicopters like the Apache are able to engage specific targets with their 30 mm chain gun, 2.75 inch rockets, and hellfire missiles. Because of their infrared gun cameras, Apaches are excellent nighttime observation platforms, providing situational awareness and pinpoint fires for ground elements.

An important feature of all rotary-wing aircraft is their ability to hover. A hovering aircraft can maintain its situational awareness and stay focused on a specific target while a high-performance aircraft cannot. Additionally, helicopters can adjust their elevation, allowing them to get below a city's skyline and more accurately engage targets. While a plane can only drop a bomb through a building's roof, the Apache can engage specific rooms on specific floors of that building.

Even though the Apache has an armored cockpit, all helicopters are extremely vulnerable from ground fire. All it takes is a single RPG round or a burst of fire from a machine-gun to hit a helicopter's main or tail rotor, sending it plummeting to the ground.

Chapter 5 – Aircraft in the Cities

OH-58 Kiowa

Photo by DOD

Because the Kiowa is an observation helicopter as opposed to a heavy attack helicopter like the Apache, it is smaller and more maneuverable. Consequently, the Kiowa can maneuver in tighter spots, land on small landing zones, and overall has greater freedom of movement in a city.

The Kiowa is more than just an observation platform and it can be configured with a variety of weapons systems to include forward firing .50 machine guns, 2.75 mm rockets, 7.62 mm mini-guns, and Hellfire missiles.

A close look at the Kiowa reveals that it is much more vulnerable than the Apache. The Kiowa has an open cockpit that means anyone with an AK-47 can shoot the pilots and bring down the entire aircraft.

Chapter 5 – Aircraft in the Cities

AH-6 "Little Bird"

As compared to its bigger brothers, like the Kiowa and Apache, the Little Bird is an ideal aerial urban warfare platform. The Little Bird too can be fitted with a variety of offensive weapons to include 2.75 inch rockets, 7.62 mm mini-guns, Hellfire anti-armor missiles, 40 mm automatic grenade launchers and .50 machine guns.

The Little Bird's biggest advantage is its exceptionally small size, allowing it to maneuver in the tightest spaces in a city. The Little Bird can fly down wide streets, land on rooftops, land at intersections, and in general has greater access to the urban environment because of the smaller diameter of its rotor blades.

The AH-6 can also be modified by attaching three bench seats on each side of the frame, allowing it to carry a total of six assault troops. This unusual modification combined with the AH-6's smaller size enables it to deliver and extract assault troops from the roofs of buildings and anywhere the Little Bird can land.

Chapter 5 – Aircraft in the Cities

Aerial Platform Shooting

Photo by DOD

Because helicopters have an excellent view of the urban terrain and are able to fly an elliptical flight track, their door gunners are extremely valuable for identifying and engaging specific targets. Door gunners can accurately engage targets at night when equipped with night vision goggles and infrared aiming lasers for their weapons systems.

This door gunner is firing a mini-gun that shoots a 7.62 x 51 mm round, a round that enables the gunner to penetrate cars, windows, and doors without overpenetrating walls or roofs. The mini-gun is electrically powered and can fire 3000 rounds a minute.

While this door gunner is firing a mini-gun, many weapon systems can be fired, using the helicopter as an aerial weapons platform. Aerial sniper teams can also be used in an urban environment, allowing a single precision shot to be made from a birds-eye perspective from a hovering helicopter.

Chapter 5 – Aircraft in the Cities

MEDEVACs

Photo by SSG Jose Hernandez

Medical evacuation (MEDEVAC) helicopters are an excellent tool for rapidly removing seriously injured personnel from the urban battlefield to a secure location. The biggest challenge for ground units requiring MEDEVAC support is locating and securing an appropriate landing zone, which are limited for a larger helicopter like the UH-60 Blackhawk shown above.

To make the MEDEVAC process easier, ground troops should identify and conduct reconnaissance of possible landing zones prior to conducting a specific mission. If the immediate area does not support a MEDEVAC, a short movement to an appropriate landing zone (like a soccer field, parking lot, or playground) will have to take place. Most MEDEVACs in an urban environment will likely be this type of ground/aerial effort due to the size of the MEDEVAC bird and the limited places to land.

Chapter 5 – Aircraft in the Cities

Urban Landing Zones

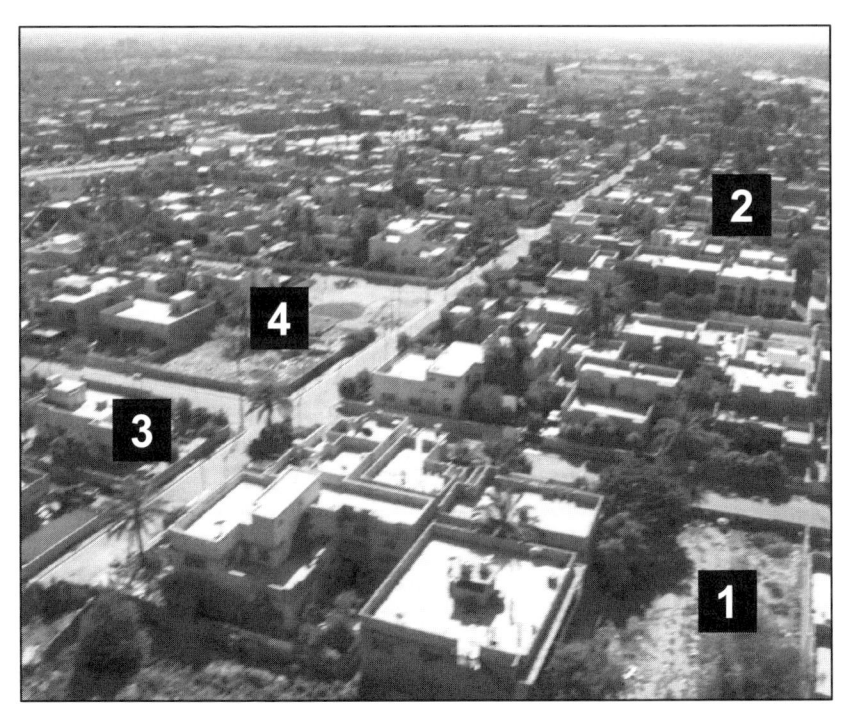

Depending on the size of the helicopter, every urban area has specific places within it that are suitable as landing zones. Big cities may have appropriate urban landing zones in the form of parks, playgrounds, soccer fields, stadiums, car parking lots, isolated farms, major intersections, wide multi-lane highways, and any building with a flat roof.

#1 is an example of an open yard.
#2 is an example of a roof top.
#3 is an example of a street intersection.
#4 is an example of an open lot.

Chapter 5 – Aircraft in the Cities

Fast-Roping Infiltration

Photo by SGT James Goff

Military organizations that are capable of conducting fast-rope operations can turn the roof of any building into an urban infiltration point. All a helicopter has to do is hover for several seconds above a selected structure, allowing a small team to slide down a rope and onto the building's roof.

Once on the roof, the assault team can conduct a vertical, top down attack, bypassing ground level doors and windows. Vertical assaults are an excellent tactic when enemy forces are in the building but are expecting a conventional ground floor entry. Few people are prepared to repel a rooftop assault.

Fast-roping is an inherently dangerous method of infiltration, complicated by nighttime conditions, sandstorms, dirt induced "brownouts", and the soldiers' own heavy equipment.

Chapter 5 – Aircraft in the Cities

Aerial Extraction

Photo by Shane Cuomo

Aerial extraction with ladders is an option for a team requiring a rooftop exit from an urban area. As long as a helicopter can hover for a period time over a selected structure, an infantry team can make an exit.

This method is time consuming, entails some risk due to the chance of potentially life-ending falls, and should not be conducted if under enemy fire. While fast-roping onto a structure is relatively easy due to gravity, climbing an extraction ladder is fatiguing and takes practice.

Chapter 5 – Aircraft in the Cities

The Risk of Aerial Ambush

Photo by SSG Suzanne Day

Because the urban environment has such a volume and variety of structures, there are thousands of possible ambush sites for enemy ground forces in any city. A helicopter crew cannot possibly identify, assess, and take defensive measures for each and every potential threat that exists in a city.

Hovering helicopters are particularly vulnerable in the cities and are easy targets for RPG gunners. The best defense for helicopters in the cities is flying low and fast, which is in itself dangerous due to any number of urban obstacles like telephone poles, water and radio towers, unseen wires, and updrafts.

Just think how easy it would be for a guerrilla armed with a shoulder-fired anti-aircraft missile to step onto a roof, fire off a single missile, and then walk back inside the structure.

By April 2006, 48 American military helicopters had been downed, at least 26 by hostile fire, with 143 fatalities.
- Brookings Institution Iraq Index

Chapter 6
Close Quarters Battle (CQB)

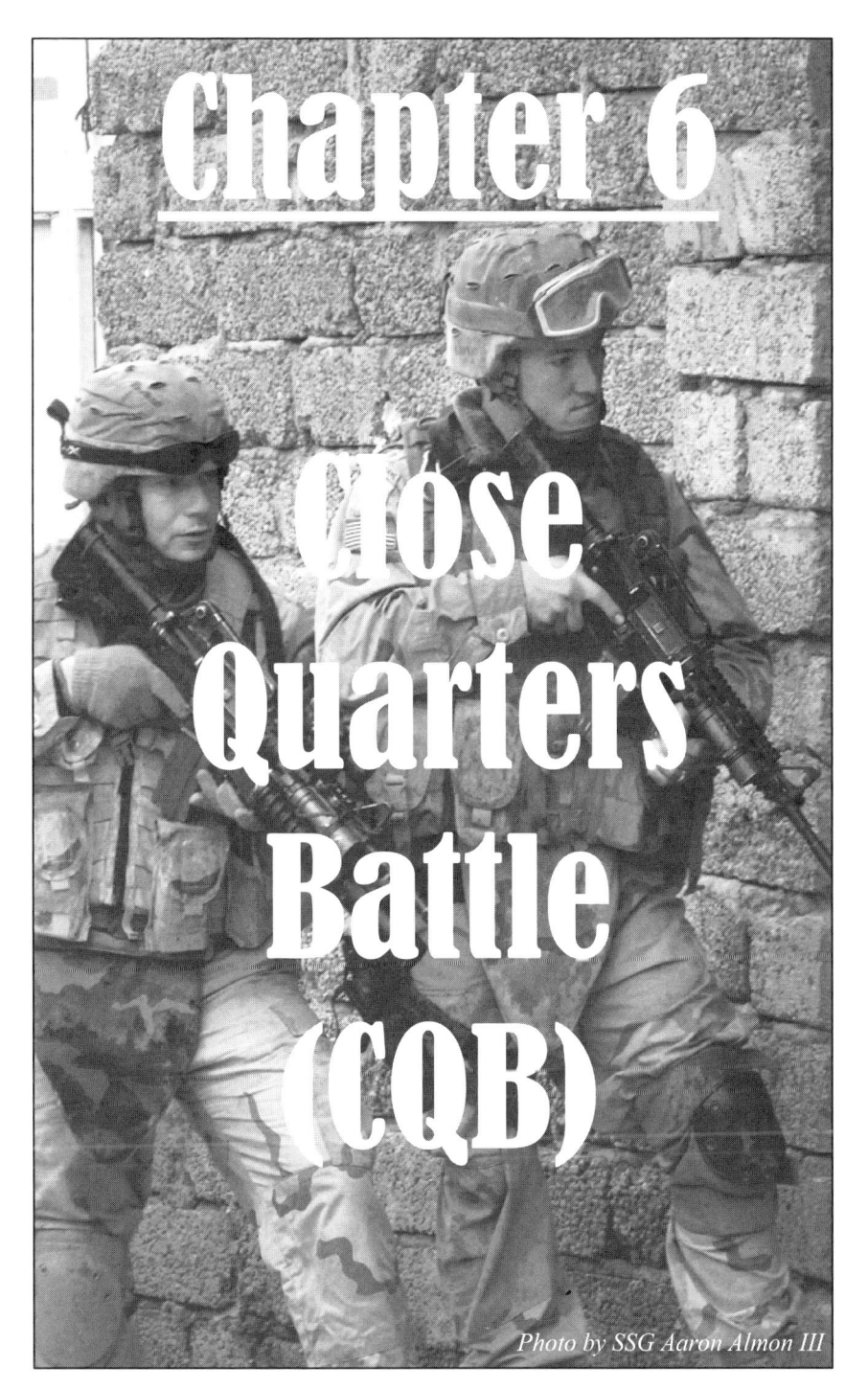

Photo by SSG Aaron Almon III

Chapter 6 – Close Quarters Battle
A Three Dimensional World

Photo by SGT Jeremiah Johnson

Unlike a rural environment, with only the two dimensional concerns of width and depth, soldiers in the cities have to operate in a three dimensional world. Potential enemy threats can come from any number of directions and elevations, both above and below a soldier.

Not only do the soldiers shown above have to be concerned with potential threats located at ground level, they must be concerned with rooftops and every level in the surrounding structures from the roof down to street level.

If a soldier is on the second story of a three story building, they must be concerned with the room above them, the room beneath them, and the rooms located on all sides of them. A three dimensional world is a complex world.

Chapter 6 – Close Quarters Battle

Rapid Environment Change
Compression to Expansion

Photo by SSG Jorge Rodriguez

The above soldiers are experiencing a rapid change in their environment. They are currently in a very compressed environment - a narrow stairwell with 360 degree cover and conccalmcnt.

As soon as they step out of this compressed microenvironment onto the roof, they transition immediately to an uncompressed macrocnvironment. One second they are concerned with a single possible threat only ten feet away and the next second they have to be concerned with a thousand possible threats a thousand feet away and everything in between.

Soldiers regularly experience this operational and physical expansion when they move out of a hallway and enter and clear a room. This environmental expansion can cause sensory overload, a phenomenon only moderated through realistic training.

Chapter 6 – Close Quarters Battle

Rapid Environment Change
Expansion to Compression

Photo by SSG James Harper

These soldiers are experiencing a different kind of rapid change in their environment. Moments before they left a relatively open macroenvironment composed of streets, intersections, structures in depth, and open sky. On the open streets they were concerned with many threats, both long distance and up close, and everything in between.

Now that these soldiers are descending stairs into a hallway, their world has become instantly compressed. Their environment has been simplified and they are now concerned with only several possible threats, but at much closer range. This very real, physical compression forces units to work in close quarters and limits the number of personnel that can operate freely in the confined space. This is another example of the urban environment forcing units to work in small-sized elements.

Chapter 6 – Close Quarters Battle

A Compressed Environment

Photo by SPC Gul Alisan

In every aspect, the urban environment is a physically compressed one. Soldiers have to move through an endless array of tight spaces such as moving through doorways and windows and they must search in closets, look under beds, crawl in tunnels, and move around the claustrophobic confines of armored vehicles.

A soldier's helmet, body armor, and equipment make the soldier less able to maneuver in these confined spaces. Soldiers must keep their personal load as streamlined as possible and they require weapons that are equally maneuverable. An overburdened soldier will not be able to move quickly through or traverse the urban environment composed of an infinite array of obstacles and barriers.

Chapter 6 – Close Quarters Battle

Body Armor

Photo by CPL Paula Fitzgerald

Photo by CPL Tom Sloan

(This Marine's life was saved by his ceramic body armor when he was shot while on patrol in Ramadi.)

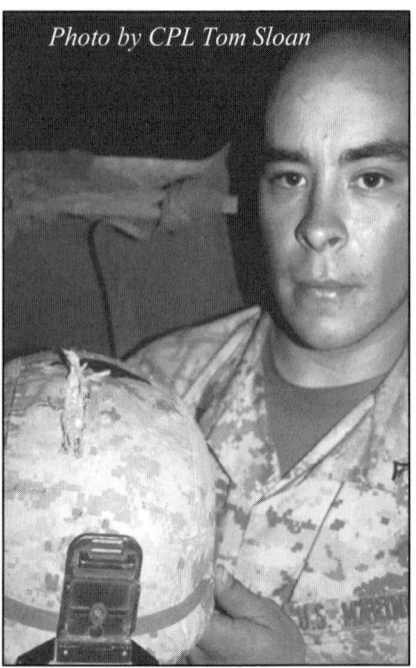

(This Marine's life was saved by his Kevlar helmet when he was shot while on patrol in Ramadi.)

Body armor is part of the urban warfare experience because, unlike the rural environment, soldiers do not have to travel long distances while on foot. Body armor is also needed to limit the higher number of casualties that are taken in the cities.

Body armor helps soldiers survive the myriad threats the urban battlefield offers like IEDs, car bombs, ambushes, shrapnel, ricochets, high-explosive blast pressure, objects thrown from rioters, car accidents, and collapsing buildings.

Body armor makes the job of enemy snipers very difficult because these snipers can no longer make a relatively easy shot to the chest. Snipers have to focus on the much more difficult head shot, which too may be thwarted by a Kevlar helmet.

Chapter 6 – Close Quarters Battle

Communications Intense

Photo by CPL Tom Sloan

Visual communication is often impossible in the urban environment because soldiers regularly experience relative isolation from being in a separate microenvironments. These small, isolated teams of soldiers still have a need to communicate – in fact they need to communicate even more except through non-visual means.

Consequently, the individual soldier requires a personal communications system. In the rural environment an organization may get away with only squad leaders having radios. In the cities each soldier must have the capability to communicate with their immediate leadership.

Chapter 6 – Close Quarters Battle

Flash-Bangs

Photo by SSG Jesus Lora

Every soldier should carry flash-bangs with them while conducting urban operations. Flash-bangs are useful as a distraction device and to disorient people when entering and clearing structures because they produce such a loud noise and a bright flash.

Flash-bangs are ideal for an urban environment because they are non-lethal, producing no fragmentation. Flash bangs are inherently safer to use because they do not penetrate walls or doors like a fragmentation grenade might. Additionally, flash-bangs produce almost know collateral infrastructure damage – at most there may be some smoke damage or burn marks. Consequently, flash-bangs can be used in close proximity to other soldiers and the civilian populace without fear of injury.

Chapter 6 – Close Quarters Battle

Weapons Camouflage

Photo by CPL Randy Bernard

In the urban environment camouflage is very important. Since there are more people present in the cities, soldiers find themselves under more scrutiny - even in battle. It is more difficult to blend into an environment dominated by concrete structures that make out of place objects and inappropriate colors stand out.

The above picture is an example of the level of detail required to camouflage individual weapons. For some reason most soldiers go to great lengths to camouflage their person but then carry around a black rifle that sticks out like a sore thumb.

The Marine shown here has painted his entire weapon with a sand colored, removable paint (probably commercially purchased Bow Flage). Additionally, he has covered his black telescopic sight so that only a small, darkened part of the lens is visible. Whatever color the dominant terrain is, soldiers' weapons need to be painted these same colors.

Chapter 6 – Close Quarters Battle

Urban Movement

The most secure way to move in the cities is through the buildings themselves, going from structure to structure. In this manner, soldiers have 360 degree cover and concealment from the outside.

Photo by SPC Teddy Wade

Photo by TSGT Andy Dunaway

Movement in a city often requires soldiers to rush from one covered and concealed position, across open areas, and into another covered and concealed position.

Because many buildings in Iraq's are built close together or share adjoining walls, soldiers have the option to move from one building to another, traveling rooftop to rooftop. By moving in this manner, soldiers can conduct top down assaults of adjoining structures.

Photo by SPC Teddy Wade

Chapter 6 – Close Quarters Battle

Killing Zones: Streets

Photo by SGT April Johnson

There is absolutely no cover and concealment in the middle of an open street. There is no way to blend into an open street and the soldiers above are clearly silhouetted against the surrounding terrain. Bullets and fragments will ricochet of the street and surrounding walls, making an open street a killing zone.

A common insurgent tactic is to ambush security forces as they move down a street confined by adjacent walls. The insurgents then close off the nearby intersections with direct fire, turning the open street in a linear ambush zone.

Soldiers should only be on an open street if they are quickly moving across it to a covered structure. If soldiers have to establish a checkpoint in a street like the one above, then the soldiers need to import cover such as driving in an armored vehicle or building a sandbagged position. If soldiers need to move in the same direction as a street, they are more secure traveling rooftop to rooftop, or moving parallel to the road, going from structure to structure, behind the adjacent walls.

Chapter 6 – Close Quarters Battle

Intersections

Photo by SSG James Harper Jr

Street intersections are dangerous places to be because they compound the danger of just a single street. As with a single street, there is no cover or concealment at an intersection. Since soldiers in a street intersection can be observed and shot at from multiple directions, soldiers need to move quickly through intersections, minimizing their exposure. Ideally, soldiers should bypass an intersection by moving into the adjacent buildings and then crossing a single street.

Street intersections attract attention because they are key terrain, facilitating the movement of both vehicles and men. A small military force can block off a relatively large urban area by controlling a few key intersections. This same military force can shut down an enemy mechanized/motorized force with a few, key placed anti-armor weapons. Importantly, infantry soldiers can climb over walls and buildings and bypass these same intersections.

Chapter 6 – Close Quarters Battle

Street Cover

Photo by SSG Klaus Baesu

The soldier shown above is making the best of a bad situation by getting into the prone position to reduce his profile. Also, this soldier is using his armored vehicle for cover as he scans for the enemy. Because his vehicle is positioned close to an adjacent wall, the soldier has minimized his exposure. The only way this soldier could improve his position is to get into a structure or drive off in his vehicle to a position with better cover.

Photo by SSG Bryan Reed

The soldiers above are in a bad position, caught in an open street, and at an intersection at that. To make matters worse, they have made the mistake of trying to get cover behind the concrete base of a streetlight while remaining in the middle of the street. This is poor cover and should only be a momentary position at best. An enemy sniper in an elevated position could easily pick off every one of these men. To solve this problem of exposure, the above soldiers could easily jump over the adjacent wall and have excellent cover and concealment.

Chapter 6 – Close Quarters Battle

Street Clutter

Photo by SPC Timothy Kingston

The above picture of a soldier providing security for a temporary checkpoint illustrates several points. First, the soldier has good observation and fields of fire down the street and can engage targets hundreds of meters away. From an enemy perspective, there is nowhere to hide in the middle of this street and the safest thing to do is get off of it.

This soldier has a good position to observe the street as he is in the prone and has some cover from the surrounding rubble and the base of this streetlight. If he were to take fire, he too would need to quickly move off the street.

This soldier has a difficult job. The surrounding trash, rubble, buildings, cars, and people create a cluttered view. This soldier has many separate activities to observe, both near and far. The hazy atmosphere obscures the people in the distance, making it difficult to determine their activities or intentions.

Chapter 6 – Close Quarters Battle

Street Corners

Photo by CPL Tom Sloan

This Marine is using the corner of a building for cover as he observes a street. The proper exploitation of corners allows a soldier to expose only a small part of their body while the soldier is still able to observe and place fires on a large area.

Photo by SSG Gregory Funk

A soldier never knows if they will have to shoot around a corner with their dominant hand or weak side hand. Therefore, soldiers must be able to shoot accurately with either hand. This is not only true for clearing corners but when clearing rooms, stairwells, small areas like closets, and unusual situations like searching under beds and vehicles.

Chapter 6 – Close Quarters Battle
Urban Climbing

Photo by GSGT Keith Milks

Photo by SPC Jerry Combes

The urban environment consists of thousands of obstacles. Locked or booby trapped doors means soldiers may have to go vertical to overcome obstacles, gain entrance to a structure, or reach higher ground for better observation and fields of fire.

The urban environment, with its ladders, door frames, ledges, and window sills lends itself to vertical assaults of some sort. Soldiers must train in urban climbing techniques so that they are prepared to traverse the urban landscape. Once a unit trains in urban climbing techniques and using basic climbing equipment, there are few structures they cannot traverse.

Chapter 6 – Close Quarters Battle

Overcoming Obstacles

Photo by SSG Klaus Baesu

Photo by SPC Timothy Story

The city is one large obstacle made up of an endless variety of individual obstacles. Some obstacles are easier to overcome than others. In Iraq almost every structure is surrounded by a courtyard wall of varying heights. Some walls are only five or six feet high while others may be over ten feet in height. Most walls, even tall ones, can be scaled with the help of a buddy and a short length of rope or tubular nylon.

The above pictures show the importance of being mobile and not over-encumbered with too much equipment. The bulkier soldiers are and the more they weigh, the harder it is to climb. A soldier's equipment must be as streamlined as possible to remain mobile in an urban environment. These pictures also illustrate the limitations of vehicles in an urban environment. Only dismounted infantry are flexible enough to traverse the full spectrum of urban obstacles.

Chapter 6 – Close Quarters Battle

The Advantage of Elevation

Photo by SGT Jeremiah Johnson

In the three dimensional world of the cities elevation is an important element of urban warfare. A tall building, because it allows a soldier to observe and place fires on every other building around it, is a key piece of terrain.

Elevation is important because it enables a soldier to look down on an urban area, similar to the bird's eye view provided by an aircraft. This top-down perspective allows an unrestricted view of the surrounding terrain.

From street level, a soldier has an extremely obstructed view. This obstruction is caused from virtually everything to include other people, cars, walls, buildings, and city blocks. Increased elevation equals better observation and better fields of fire.

Chapter 6 – Close Quarters Battle

An Elevated Perspective

Photo by SGT Jeremiah Johnson

This picture illustrates the power of elevation. A soldier in an elevated position can look down on a street without effort. However, it is unnatural and therefore fatiguing for street level soldiers to constantly look upwards. Consequently, soldiers in elevated positions enjoy inherently better cover and concealment as compared to street level soldiers.

From the elevated perspective of the photographer taking this picture, one can see several advantages of elevation. First, an enemy soldier from the roof could shoot the Stryker armored vehicle on its roof with an RPG, bypassing its stronger side armor. Depending on the Stryker's weapons systems, the vehicle may not be able to elevate its weapon high enough to target the enemy soldier. Additionally, the accompanying foot patrol has no place to hide and could be easily picked off with sniper fire. If the infantry in this picture try and hide behind the vehicle, a sniper may still be able to pick them off due to the sniper's elevated angle.

Chapter 6 – Close Quarters Battle

Rooftop Positions

Photo by LCPL Graham Paulsgrove

By securing a rooftop position, these Marines have gained the high ground, ensuring good observation and fields of fire of the surrounding area. They have minimized their silhouettes by staying low with just their helmets and weapons exposed.

A person trying to shoot from street level up at them simply does not have a good enough angle to hit them. Their rounds will either hit the building or travel over the two Marines' heads. In contrast, the two Marines can look down into the streets and see and hit everything.

If one looks closely at this picture, they will notice how these Marines' black weapons stick out as compared to the tan color of their uniforms and the surrounding structure. Their weapons, to include their sights, should be spray painted to match the surrounding terrain.

Chapter 6 – Close Quarters Battle

The Danger of Rooftop Positions

Photo by SGT Michael Cardin

This picture shows exactly why rooftop positions – in a high-intensity warfare environment – are dangerous places to be. While rooftop positions do give soldiers an elevated position, these same rooftop positions offer no cover from aerial bursting munitions. Above, an artillery aerial burst, used specifically to clean rooftops of enemy personnel, detonates perfectly above the building.

Not only are rooftops dangerous places to be because of air burst indirect fire, but the top floor of a structure may be equally dangerous. Artillery that does not airburst will impact on the roof, detonate, and then kill or injure everyone on the top floor. When under indirect fire barrages, it is best to be at least one floor down from the uppermost level. This way the entire top floor serves as a buffer zone, catching incoming blasts, debris, and shrapnel.

Chapter 6 – Close Quarters Battle

Breaching Techniques

Soldiers should not assume that they have to break down a door to get inside a structure – getting inside might be as easy as using the door handle.

Photo by SFC Johancharles van Boers

Depending on the design of the door and the door frame, kicking the door at the locking mechanism may be enough to force it open. This will work most of the time on interior doors that are weak by design.

Photo by TSGT Russell Cooley

Doors made of sheet metal are flexible and if the locking mechanism cannot be forced open, a corner can be pried opened, allowing soldiers to either slide through or open the door from the inside.

Photo by TSGT Andy Dunaway

Chapter 6 – Close Quarters Battle

Bypass Breaching Techniques

If a courtyard gate is locked, soldiers might not have to force it open. They can just climb over the wall, then open the gate from the inside, or bypass the gate altogether. Insurgent forces are known to place anti-personnel mines where soldiers may climb over a wall.

Photo by SPC Lester Colley

If the door to a structure is locked or barricaded, another opening into the structure, like an open window, may allow entry. Again, a soldier and their equipment must be streamlined to fit through such a tight space.

Photo by SPC Gul Alisan

Windows are always a good point for an alternate entry into a structure. However, if a window has a steel security frame, it may be as difficult to breach as a door.

Photo by LCPL Nathan Ferbert

147

Chapter 6 – Close Quarters Battle

Mechanical Breaching Techniques

Photo by LCPL Kaemmerer

Mechanical means of breaching might have to be used for steel doors that have solid steel frames and strong locking mechanisms. Bolt cutters are excellent for getting past padlocked doors.

Photo by CPL Tom Sloan

A pry bar is a useful tool allowing soldiers to force open doors, break open padlocks and bypass locking mechanisms. Most doors in the cities can be opened with this tool.

Photo by SGT Derek Gains

A battering tool is another means for breaking through doors at the locking mechanism. Only steel framed doors have a chance to withstand the blows from a heavy battering tool.

Chapter 6 – Close Quarters Battle

Breaching Techniques

Photo by SGT Jeremiah Johnson

A portable saw designed for cutting through steel may be required to breach heavy steel doors that are common in certain urban areas, such as in military facilities and industrial parks.

Photo by LCPL Paul Robbins

Photo by SGT Clinton Firstbrook

12-gauge shotguns are a fast, effective way for breaching doors by shooting either at the locking mechanism or at the hinges of the door. Unlike other breaching tools, the shotgun can also be used to shoot enemy personnel immediately after securing the breach.

Chapter 6 – Close Quarters Battle

Explosive Breaching Techniques

Photo by CPL Randy Bernard

Explosive breaching is relatively simple, only requiring ¼ pound of C4 military explosives, an electric blasting cap, and a power source to detonate the explosives to blow open any door.

Photo by SGT Clinton Firstbrook

Explosive breaching, while allowing soldiers to enter otherwise impassable doors, also disorients the people inside the structure. Explosives can be used to breach any structural medium to include walls, roofs, and floors.

Chapter 6 – Close Quarters Battle

Breaching Techniques

Photo by LCPL Will Lathrop

A vehicle can be used to breach, crush, and collapse obstacles like courtyard walls and even the walls into a structure. The collapsed wall now becomes the breach point.

Photo by DOD

The above picture shows a courtyard wall that has been breached by a an aerial guided munition – a type of explosive breach. If there is enough distance between the point of impact and surrounding structures, collateral damage is minimized.

Chapter 6 – Close Quarters Battle

Interior Doorways

Photo by SPC Jeremy Crisp

Once inside a structure, doorways become danger areas. People are expected to move through doorways as they are natural breach or entry points inside a structure.

Enemy personnel expect soldiers to move through doorways and will focus their attention on these areas. During CQB encounters, a large percentage of wounded soldiers are shot as they breach a structure or enter through a doorway.

Soldiers should never stand in or near doorways. They must pass through these entry points and then choose a better place offering some cover or at least an area where they are not silhouetted by the doorway. Standing in front of a doorway inside a house is equivalent to standing on top of a roof out on the streets.

Chapter 6 – Close Quarters Battle

Hallways

Photo by DOD

A hallway inside a building is equivalent to a street outside – they are dangerous places to be. Hallways are killing funnels, bouncing bullets, fragments, and high-explosive blasts into the people moving through them.

Soldiers need to travel quickly through a hallway, moving from room to room, minimizing their time and exposure in the hallway. If soldiers want to avoid using a hallway they can travel from room to room, paralleling the hallway, as they explosively breach their way through the walls of the structure.

Hallways are key terrain in a structure because they allow people to move rapidly through them. Therefore, soldiers need to observe and control hallways without standing exposed in them. Many CQB fights revolve around attempts to control a structure's hallway, especially if there is only one hallway in the structure connecting all of the separate rooms.

Chapter 6 – Close Quarters Battle

Stairwells

Stairwells are key terrain inside a structure. They are high speed avenues of approach allowing people to move quickly from one level to another. Whomever controls the top of a stairwell has a distinct advantage over those fighting to get up them.

Photo by Russell Cooley

Once a team gets inside a structure, they must immediately secure the stairwells. Even if the team does not clear the stairwell at first, they must lock it down and put eyes and a gun on it.

Photo by TSGT Andy Dunaway

Chapter 6 – Close Quarters Battle

Stairwells

Photo by CPL Tom Sloan

Photo by CPL Mike Escobar

It takes several trained soldiers to clear a stairwell with minimal risk. Different soldiers in a clearing team must secure different fields of view as they ascend the stairs. This is a difficult task even for trained men and enemy personnel at the top of a stairwell have a distinct advantage while the people ascending the stairs are at a disadvantage. Even untrained men, like guerrillas, can stop professionals from getting up a set of stairs.

Once a stairwell is under control, this key terrain must be secured. Therefore, it is a good policy to place a man at every stairwell until the entire structure is cleared.

If an element cannot fight its way up a flight of stairs, then they must bypass this obstacle and conduct a ceiling breach somewhere else in the structure. The soldiers can then get to the next level in the structure without using the stairs at all.

Chapter 6 – Close Quarters Battle

Stairwells

Photo by CPL Tom Sloan

Photo by CPL Thomas Grove

With narrow steps and no handrails, some stairwells are dangerous. Stairwells like these are an obstacle that is difficult to negotiate, making ascending or descending them a slow task. As with hallways, many CQB situations revolve around one element trying to fight their way up a set of stairs.

Even outside stairs force soldiers to take a specific, known path. A team of soldiers crowded on an exposed stairwell make an inviting target. An experienced defender will booby-trap stairways with explosives and cover them with direct fire from machine-guns and snipers. In this manner, a prepared defender turns the stairs into a killing zone, ambushing their enemy while they are caught in narrow, confining substructure.

Chapter 6 – Close Quarters Battle

Nighttime Lighting

During night operations a light source like a streetlamp will expose soldiers. The soldiers must either move out of the light and into the shadows or cut off the light source.

Photo by PFC Bryan Kinkade

Even if a soldier is not directly in the light, standing in front of a light source will silhouette them. A suppressed weapon, or even an air rifle, is ideal for quietly shooting out a light.

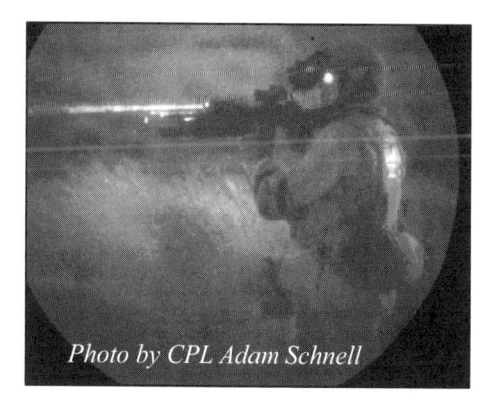

Photo by PFC Bryan Kinkade

Operating in complete darkness is more secure. American military forces have the night vision technology to operate without light. The insurgents do not have this technology and are at a distinct disadvantage during night combat operations.

Photo by CPL Adam Schnell

Chapter 6 – Close Quarters Battle

Daytime Lighting

Photo by SPC Arthur Hamilton

These soldiers are about to enter a dark room, going from a lighted environment that demands sunglasses, into an environment that is pitch black. Unless this soldier has a light, he will be temporarily blinded until his eyes adjust to the new environment.

Photo by TSGT Russell Cooley

Darkened spaces prevent effective observation into them, even in daylight, and are therefore perfect for sniper hides.

Photo by SFC Johancharles van Boers

As illustrated by this burning structure in daylight, soldiers may encounter an unpredictable variety of unusual light and shadow combinations.

Chapter 6 – Close Quarters Battle

Flashlights

The factory made, gun mounted flashlight is a mandatory tool and should be mounted 24 hours a day on a soldier's weapon. A flashlight may need to be used both during nighttime and daytime conditions.

Photo by CPL Paul Leicht

The flashlight is valuable during nighttime operations, like searching through identification documents and passports during the conduct of a raid. It is also critical for clearing rooms and identifying if an unknown person is a threat or not a threat.

Photo by CPL Tom Sloan

In the urban environment, flashlights are needed during the day. Even in daylight hours, soldiers must have the ability to project light inside a structure. If a structure is windowless, it will be as dark as night inside even if it is light outside.

Photo by PVT Brandi Marshall

Chapter 6 – Close Quarters Battle

Mistake: Skyline Silhouetting

Photo by SGT Michael Abney

Silhouetting is a constant consideration in the urban terrain and there are many different forms of silhouetting. The most common form is whenever a soldier gets above the highest point of the surrounding terrain while they are standing, kneeling, or lying down. Because they are above the surrounding terrain, they stick out against the skyline and are a target. Enemy snipers look for skyline silhouetting, both day and night.

Soldiers should never stand up on a rooftop as they have no cover or concealment and are a lucrative target from all directions. While these soldier can see and shoot in all directions, they can also be seen and be shot at from all directions.

Instead of presenting themselves on the roofs like targets, these soldiers could be inside the building they are standing on, looking out from the windows while securing themselves a position with 360 degree cover and concealment.

Chapter 6 – Close Quarters Battle

Mistake: Silhouetting

Photo by SGT Joseph Chenelly

No matter what a soldier's camouflage is or what is behind the door, anyone standing in an open doorway is instantly silhouetted. The only time a soldier should be in a doorway is if they are moving through it.

Photo by SSG James Harper Jr.

Even though this soldier is not skylined or near a door or window, he is still silhouetting himself. This soldier's green uniform and black weapon do not blend into the surrounding terrain. Whenever a soldier and their equipment do not match their background, they are silhouetted.

Chapter 6 – Close Quarters Battle

Mistake: Guard Tower Silhouetting

Photo by Jimmy Lane Jr.

This soldier's pink face, pink hands, black sunglasses, black chin strap, and black machine gun are silhouetted against the tan background. Anything failing to match the natural background of an area sticks out immediately. Enemy snipers can easily pick out these color discrepancies and there are several insurgent snipers known to be operating in Iraq who specialize in shooting guards in guard towers.

One way to defeat this silhouetting/color mismatch problem is by first placing a solid background behind the soldier, like a piece of plywood. Then, place fine mesh netting over the opening of the observation tower. With the mesh netting combined with the soldier standing in front of a solid background, people looking into the tower will only see a wall of netting. The observer behind the mesh is completely concealed but can still see out.

Chapter 6 – Close Quarters Battle

Mistake: Keyhole Shot Exposure

Photo by PFC Eric Ledrew

These soldiers need to be aware that being inside a structure alone does not automatically make them safe from enemy fire. It is human nature to become focused on one's immediate surroundings and forget that a shot can be made from outside and into a structure. However, soldiers must treat all openings as danger areas, not just windows and doorways.

The large opening in the wall behind the soldier's back is a perfect opening for enemy snipers to shoot through. In fact snipers look to make such a keyhole shot because their own position remains concealed from their target. If the uppermost soldier was shot in the back from a sniper, the lower soldier would have no idea where the shot came from, allowing the sniper to reposition undetected or even take another shot.

Chapter 6 – Close Quarters Battle

Mistake: No Cover

Photo by SGT Andre Reynolds

This soldier is sitting in an exposed location. He is in the worst possible location – in a corner. Bullets tend to hit and then skip off of walls, traveling parallel to them until they hit something solid like a corner. This soldier should move to either of the two door positions and use the shadows of the room for concealment.

Photo by SGT Jeremiah Johnson

These soldiers have absolutely no cover and concealment. While there are plenty of covered positions near their vehicle, they choose to remain in the center of an open street. If these soldiers need to protect their vehicle from enemy attacks, they can do this better by taking positions in the surrounding structures.

Chapter 6 – Close Quarters Battle

Mistake: Shallow Positioning

Photo by DOD

Interior silhouetting is dangerous. In general, windows are dangerous places to be inside a structure. The above soldiers are positioned right next to the window so that they can be seen from outside and therefore shot from outside. These soldiers need to be positioned deeper in the room, off to the sides of the window so they cannot be seen from outside.

Photo by DOD

This soldier is so far forward in his position that the barrel of his weapon is sticking out. Enemy snipers scanning the urban terrain will quickly identify and then target this soldier. This soldier should be positioned away from the opening in the wall so that he and his weapon are concealed by the shadows of the room.

Chapter 6 – Close Quarters Battle

The Snipers' War

Photo from www.dragunov.net

This insurgent sniper is sighting through a Russian made scope mounted to a 7.62 mm Tabuk rifle, a modified Iraqi version of a Yugoslavian design. The Tabuk is a reliable weapon, not as accurate as bolt-action American sniper rifles, but when used with a scope it can provide accurate fire against man-sized targets at 500 meters. The Tabuk rifle is a commonly found weapon, readily available on the black market.

One insurgent sniper, called "Juba" by Iraqi insurgent groups, is believed to be using a Tabuk rifle with a Russian scope to conduct his attacks. "Juba's" specialty is firing at his targets from inside a vehicle, in broad daylight, in dense urban areas, filming the attacks with a gun-mounted camera for later release in insurgent propaganda videos.

While many insurgents employ sniper rifles, few are actually trained professionals exploiting the capabilities of their weapons. However, several skilled insurgent sniper teams are known to work Iraq's major cities like Fallujah, Ramadi, and Baghdad.

Chapter 6 – Close Quarters Battle

Snipers: Pinning Superior Forces

Photo by LCPL Joel Chaverri

The sniper is an excellent tool, allowing one man to hold off a larger force through superior tactics and well placed fire. The Marines above are taking cover in a structure from an insurgent sniper as they advance through Fallujah.

As can be seen in this photograph, a disciplined sniper shooting from a concealed position can pin down a large number of soldiers, allowing other insurgents to maneuver against their enemy or to retreat under superior firepower.

Two snipers working in a pair can leapfrog while in the city, stalling the security forces' attack. With the leapfrog tactic, one sniper engages the enemy while the other repositions. Then the newly repositioned sniper takes aim while the other sniper repositions. The two snipers can either leapfrog away from their enemy or move laterally to hit them on their flanks.

Chapter 6 – Close Quarters Battle

Snipers in Fallujah

Photo by CPL Joel Chaverri

In this gruesome photo, an insurgent sniper catches a group of Marines in an open street during the assault on Fallujah. The insurgent sniper shot the Marine on the right first. When the other two Marines tried to pull him out of the line of fire, the sniper shot the Marine laying on the left.

A sniper shooting one soldier, waiting for help to arrive, and then shooting the next soldier is a common sniper tactic. In this case, the first wounded soldier becomes the bait for additional soldiers, who then become bait for other soldiers, and so on.

These ugly events underline the fact that an open street is a dangerous place to be in a gunfight. These Marines were shot moving down an open street, vulnerable to sniper fire from a variety of positions. It would have been safer for the Marines to travel from structure to structure (like the ones visible in the background) as opposed to risking movement in a kill zone.

Chapter 6 – Close Quarters Battle

Countersniper Tactics

Photo by LCPL Joel Chaverri

These two Marines in Fallujah display excellent countersniper skills. The Marine on the left raises his helmet above the wall, enticing an enemy sniper to shoot at it, thinking it is a Marine looking over the wall. If and when the enemy sniper shoots at the helmet, the Marine on the right, looking through the loophole, will attempt to identify the enemy sniper's position from the muzzle flash and blast from his weapon.

If the Marine does see the enemy sniper's position he may take a shot at the sniper. Because the Marine is shooting through a small hole, the enemy sniper will be unable to see the Marine's muzzle flash or blast. If the Marines do not want to give away their loophole position and do not want the enemy to know they are a countersniper team, they can simply identify the sniper's position. Once identified they can have another element deal with the threat either with an airstrike, armored vehicles, an infantry assault, or a Marine sniper team.

Chapter 6 – Close Quarters Battle

Urban Sniping Equipment

Photo by SGT Jeremiah Johnson

This sniper displays some equipment that is key for conducting urban sniper operations. His bolt-action 7.62 mm, match-grade Remington rifle is very accurate at long distances, capable of hitting man-sized targets at 600 meters and further. Notice how the rifle has been painted a sand color to help in blend into the surrounding terrain.

An AN/ PVS-10 day/night scope is mounted on the rifle. This scope is a valuable tool because it allows the sniper to observe and engage targets both day and night.

The sniper is also using an adjustable height tripod to support his weapon. The adjustable tripod is an excellent tool enabling the sniper to adjust his weapon's elevation according to the surrounding cover.

Chapter 6 – Close Quarters Battle

Sniper Positions

Photo by LCPL James Vooris

This sniper is using a .50 rifle, capable of killing enemy personnel at a distance of a mile. The .50 rifle is a defensive weapon system, is not very maneuverable, and is ideal when fired within friendly lines where displacement after the shot is not required or even desired. This Marine has a very comfortable position and can observe a specific area for extended periods of time.

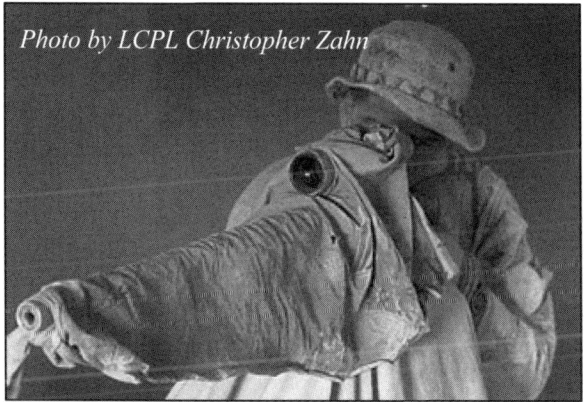

Photo by LCPL Christopher Zahn

This sniper in Fallujah has excellent urban camouflage. The material covering his weapon is smooth and featureless – it almost looks like a leather chamois cloth for drying cars. Also this sniper has an appropriately colored sheet covering his body from the shoulders down. Considering the background color and if the sniper is deep enough in this room, even his pink face blends in.

Chapter 6 – Close Quarters Battle

The Keyhole Shot

Photo by SGT Jose Guillen

This sniper in Fallujah shows excellent snipercraft by taking a keyhole shot. By shooting through this relatively small hole in the wall the sniper can observe and engage targets out at long distances. However, enemy personnel in return cannot see the sniper through the same small hole.

The wall provides the sniper cover and concealment, making him almost invulnerable to enemy direct fire. The wall will also hide the sniper's muzzle flash and muzzle blast when he fires. As result, this sniper will most likely remain invisible to his targets, even after making the shot, because his weapon's muzzle blast and flash is what the enemy looks for.

Because of all these advantages, the keyhole shot is the preferred shot for the urban sniper. A sniper who can make a shot undetected does not necessarily have to reposition after their shot to avoid exposure. They may be able to take a shot, lay low for a few hours, and then take another shot.

All information for this work can be found in the public domain. An excellent source for further information on the subject can be found at www.wikipedia.com, which has detailed information on the Iraqi insurgency and urban warfare.

Also, www.defenselink.mil has thousands of relevant pictures with accompanying stories about the current war in Iraq.

The excellent work *No True Glory* by Bing West describes in detail modern urban warfare, both high and low-intensity, in Fallujah between conventional and guerrilla forces.